U·X·L
Encyclopedia
of Science

U·X·L
Encyclopedia
of Science

Second Edition

Volume 7: Mas-O

Rob Nagel, Editor

FARMINGDALE PUBLIC LIBRARY
116 MERRITTS ROAD
FARMINGDALE, N.Y. 11735

GALE GROUP

THOMSON LEARNING

Detroit • New York • San Diego • San Francisco
Boston • New Haven, Conn. • Waterville, Maine
London • Munich

U•X•L Encyclopedia of Science
Second Edition

Rob Nagel, *Editor*

Staff

Elizabeth Shaw Grunow, *U•X•L Editor*

Julie Carnagie, *Contributing Editor*

Carol DeKane Nagel, *U•X•L Managing Editor*

Thomas L. Romig, *U•X•L Publisher*

Shalice Shah-Caldwell, *Permissions Associate (Pictures)*

Robyn Young, *Imaging and Multimedia Content Editor*

Rita Wimberley, *Senior Buyer*

Pamela A. E. Galbreath, *Senior Art Designer*

Michelle Cadorée, *Indexing*

GGS Information Services, *Typesetting*

On the front cover: Nikola Tesla with one of his generators, reproduced by permission of the Granger Collection.

On the back cover: The flow of red blood cells through blood vessels, reproduced by permission of Phototake.

Library of Congress Cataloging-in-Publication Data

U-X-L encyclopedia of science.—2nd ed. / Rob Nagel, editor
p.cm.
Includes bibliographical references and indexes.
Contents: v.1. A-As — v.2. At-Car — v.3. Cat-Cy — v.4. D-Em — v.5. En-G — v.6. H-Mar — v.7. Mas-O — v.8. P-Ra — v.9. Re-St — v.10. Su-Z.
Summary: Includes 600 topics in the life, earth, and physical sciences as well as in engineering, technology, math, environmental science, and psychology.
ISBN 0-7876-5432-9 (set : acid-free paper) — ISBN 0-7876-5433-7 (v.1 : acid-free paper) — ISBN 0-7876-5434-5 (v.2 : acid-free paper) — ISBN 0-7876-5435-3 (v.3 : acid-free paper) — ISBN 0-7876-5436-1 (v.4 : acid-free paper) — ISBN 0-7876-5437-X (v.5 : acid-free paper) — ISBN 0-7876-5438-8 (v.6 : acid-free paper) — ISBN 0-7876-5439-6 (v.7 : acid-free paper) — ISBN 0-7876-5440-X (v.8 : acid-free paper) — ISBN 0-7876-5441-8 (v.9 : acid-free paper) — ISBN 0-7876-5775-1 (v.10 : acid-free paper)

1. Science-Encyclopedias, Juvenile. 2. Technology-Encyclopedias, Juvenile. [1. Science-Encyclopedias. 2. Technology-Encyclopedias.] I. Title: UXL encyclopedia of science. II. Nagel, Rob.
Q121.U18 2001
503-dc21

2001035562

This publication is a creative work fully protected by all applicable copyright laws, as well as by misappropriation, trade secret, unfair competition, and other applicable laws. The editors of this work have added value to the underlying factual material herein through one or more of the following: unique and original selection, coordination, expression, arrangement, and classification of the information. All rights to this publication will be vigorously defended.

Copyright © 2002   U•X•L, an imprint of The Gale Group

All rights reserved, including the right of reproduction in whole or in part in any form.

Printed in the United States of America

10 9 8 7 6 5 4 3 2 1

Table of Contents

Reader's Guide vii

Entries by Scientific Field ix

Volume 1: A–As 1
    Where to Learn More xxxi
    Index xxxv

Volume 2: At–Car 211
    Where to Learn More xxxi
    Index xxxv

Volume 3: Cat–Cy 413
    Where to Learn More xxxi
    Index xxxv

Volume 4: D–Em 611
    Where to Learn More xxxi
    Index xxxv

Volume 5: En–G 793
    Where to Learn More xxxi
    Index xxxv

Volume 6: H–Mar 1027
    Where to Learn More xxxi
    Index xxxv

Volume 7: Mas–O 1235
    Where to Learn More xxxi
    Index xxxv

Contents

Volume 8: P-Ra 1457
    Where to Learn More xxxi
    Index xxxv

Volume 9: Re-St 1647
    Where to Learn More xxxi
    Index xxxv

Volume 10: Su-Z 1829
    Where to Learn More xxxi
    Index xxxv

Reader's Guide

Demystify scientific theories, controversies, discoveries, and phenomena with the *U•X•L Encyclopedia of Science,* Second Edition.

This alphabetically organized ten-volume set opens up the entire world of science in clear, nontechnical language. More than 600 entries—an increase of more than 10 percent from the first edition—provide fascinating facts covering the entire spectrum of science. This second edition features more than 50 new entries and more than 100 updated entries. These informative essays range from 250 to 2,500 words, many of which include helpful sidebar boxes that highlight fascinating facts and phenomena. Topics profiled are related to the physical, life, and earth sciences, as well as to math, psychology, engineering, technology, and the environment.

In addition to solid information, the *Encyclopedia* also provides these features:

- "Words to Know" boxes that define commonly used terms
- Extensive cross references that lead directly to related entries
- A table of contents by scientific field that organizes the entries
- More than 600 color and black-and-white photos and technical drawings
- Sources for further study, including books, magazines, and Web sites

Each volume concludes with a cumulative subject index, making it easy to locate quickly the theories, people, objects, and inventions discussed throughout the *U•X•L Encyclopedia of Science,* Second Edition.

Reader's Guide

Suggestions

We welcome any comments on this work and suggestions for entries to feature in future editions of *U•X•L Encyclopedia of Science*. Please write: Editors, *U•X•L Encyclopedia of Science,* U•X•L, Gale Group, 27500 Drake Road, Farmington Hills, Michigan, 48331-3535; call toll-free: 800-877-4253; fax to: 248-699-8097; or send an e-mail via www.galegroup.com.

Entries by Scientific Field

Boldface indicates volume numbers.

Acoustics

Acoustics	**1**:17
Compact disc	**3**:531
Diffraction	**4**:648
Echolocation	**4**:720
Magnetic recording/ audiocassette	**6**:1209
Sonar	**9**:1770
Ultrasonics	**10**:1941
Video recording	**10**:1968

Aerodynamics

Aerodynamics	**1**:39
Fluid dynamics	**5**:882

Aeronautical engineering

Aircraft	**1**:74
Atmosphere observation	**2**:215
Balloon	**1**:261
Jet engine	**6**:1143
Rockets and missiles	**9**:1693

Aerospace engineering

International Ultraviolet Explorer	**6**:1120
Rockets and missiles	**9**:1693
Satellite	**9**:1707
Spacecraft, manned	**9**:1777
Space probe	**9**:1783
Space station, international	**9**:1788
Telescope	**10**:1869

Agriculture

Agriculture	**1**:62
Agrochemical	**1**:65
Aquaculture	**1**:166
Biotechnology	**2**:309
Cotton	**3**:577
Crops	**3**:582
DDT (dichlorodiphenyl-trichloroethane)	**4**:619
Drift net	**4**:680
Forestry	**5**:901
Genetic engineering	**5**:973
Organic farming	**7**:1431
Slash-and-burn agriculture	**9**:1743
Soil	**9**:1758

Anatomy and physiology

Anatomy	**1**:138
Blood	**2**:326

U·X·L Encyclopedia of Science, 2nd Edition

Entries by Scientific Field

Brain	**2:**337
Cholesterol	**3:**469
Chromosome	**3:**472
Circulatory system	**3:**480
Digestive system	**4:**653
Ear	**4:**693
Endocrine system	**5:**796
Excretory system	**5:**839
Eye	**5:**848
Heart	**6:**1037
Human Genome Project	**6:**1060
Immune system	**6:**1082
Integumentary system	**6:**1109
Lymphatic system	**6:**1198
Muscular system	**7:**1309
Nervous system	**7:**1333
Physiology	**8:**1516
Reproductive system	**9:**1667
Respiratory system	**9:**1677
Skeletal system	**9:**1739
Smell	**9:**1750
Speech	**9:**1796
Taste	**10:**1861
Touch	**10:**1903

Anesthesiology

Alternative medicine	**1:**118
Anesthesia	**1:**142

Animal husbandry

Agrochemical	**1:**65
Biotechnology	**2:**309
Crops	**3:**582
Genetic engineering	**5:**973
Organic farming	**7:**1431

Anthropology

Archaeoastronomy	**1:**171
Dating techniques	**4:**616
Forensic science	**5:**898
Gerontology	**5:**999
Human evolution	**6:**1054
Mounds, earthen	**7:**1298
Petroglyphs and pictographs	**8:**1491

Aquaculture

Aquaculture	**1:**166
Crops	**3:**582
Drift net	**4:**680
Fish	**5:**875

Archaeology

Archaeoastronomy	**1:**171
Archaeology	**1:**173
Dating techniques	**4:**616
Fossil and fossilization	**5:**917
Half-life	**6:**1027
Nautical archaeology	**7:**1323
Petroglyphs and pictographs	**8:**1491

Artificial intelligence

Artificial intelligence	**1:**188
Automation	**2:**242

Astronomy

Archaeoastronomy	**1:**171
Asteroid	**1:**200
Astrophysics	**1:**207
Big bang theory	**2:**273
Binary star	**2:**276
Black hole	**2:**322
Brown dwarf	**2:**358
Calendar	**2:**372
Celestial mechanics	**3:**423
Comet	**3:**527
Constellation	**3:**558
Cosmic ray	**3:**571

Entries by Scientific Field

Cosmology	3:574	Sun	10:1844
Dark matter	4:613	Supernova	10:1852
Earth (planet)	4:698	Telescope	10:1869
Eclipse	4:723	Ultraviolet astronomy	10:1943
Extrasolar planet	5:847	Uranus (planet)	10:1952
Galaxy	5:941	Variable star	10:1963
Gamma ray	5:949	Venus (planet)	10:1964
Gamma-ray burst	5:952	White dwarf	10:2027
Gravity and gravitation	5:1012	X-ray astronomy	10:2038
Infrared astronomy	6:1100		
International Ultraviolet Explorer	6:1120	## Astrophysics	
Interstellar matter	6:1130	Astrophysics	1:207
Jupiter (planet)	6:1146	Big bang theory	2:273
Light-year	6:1190	Binary star	2:276
Mars (planet)	6:1228	Black hole	2:322
Mercury (planet)	7:1250	Brown dwarf	2:358
Meteor and meteorite	7:1262	Celestial mechanics	3:423
Moon	7:1294	Cosmic ray	3:571
Nebula	7:1327	Cosmology	3:574
Neptune (planet)	7:1330	Dark matter	4:613
Neutron star	7:1339	Galaxy	5:941
Nova	7:1359	Gamma ray	5:949
Orbit	7:1426	Gamma-ray burst	5:952
Pluto (planet)	8:1539	Gravity and gravitation	5:1012
Quasar	8:1609	Infrared astronomy	6:1100
Radio astronomy	8:1633	International Ultraviolet Explorer	6:1120
Red giant	9:1653	Interstellar matter	6:1130
Redshift	9:1654	Light-year	6:1190
Satellite	9:1707	Neutron star	7:1339
Saturn (planet)	9:1708	Orbit	7:1426
Seasons	9:1726	Quasar	8:1609
Solar system	9:1762	Radio astronomy	8:1633
Space	9:1776	Red giant	9:1653
Spacecraft, manned	9:1777	Redshift	9:1654
Space probe	9:1783	Space	9:1776
Space station, international	9:1788	Star	9:1801
Star	9:1801	Starburst galaxy	9:1806
Starburst galaxy	9:1806	Star cluster	9:1808
Star cluster	9:1808	Stellar magnetic fields	9:1820
Stellar magnetic fields	9:1820	Sun	10:1844

U·X·L Encyclopedia of Science, 2nd Edition   xi

Entries by Scientific Field

Supernova	**10:**1852
Ultraviolet astronomy	**10:**1943
Uranus (planet)	**10:**1952
Variable star	**10:**1963
White dwarf	**10:**2027
X-ray astronomy	**10:**2038

Atomic/Nuclear physics

Actinides	**1:**23
Alkali metals	**1:**99
Alkali earth metals	**1:**102
Alternative energy sources	**1:**111
Antiparticle	**1:**163
Atom	**2:**226
Atomic mass	**2:**229
Atomic theory	**2:**232
Chemical bond	**3:**453
Dating techniques	**4:**616
Electron	**4:**768
Half-life	**6:**1027
Ionization	**6:**1135
Isotope	**6:**1141
Lanthanides	**6:**1163
Mole (measurement)	**7:**1282
Molecule	**7:**1285
Neutron	**7:**1337
Noble gases	**7:**1349
Nuclear fission	**7:**1361
Nuclear fusion	**7:**1366
Nuclear medicine	**7:**1372
Nuclear power	**7:**1374
Nuclear weapons	**7:**1381
Particle accelerators	**8:**1475
Quantum mechanics	**8:**1607
Radiation	**8:**1619
Radiation exposure	**8:**1621
Radiology	**8:**1637
Subatomic particles	**10:**1829
X ray	**10:**2033

Automotive engineering

Automobile	**2:**245
Diesel engine	**4:**646
Internal-combustion engine	**6:**1117

Bacteriology

Bacteria	**2:**253
Biological warfare	**2:**287
Disease	**4:**669
Legionnaire's disease	**6:**1179

Ballistics

Ballistics	**2:**260
Nuclear weapons	**7:**1381
Rockets and missiles	**9:**1693

Biochemistry

Amino acid	**1:**130
Biochemistry	**2:**279
Carbohydrate	**2:**387
Cell	**3:**428
Cholesterol	**3:**469
Enzyme	**5:**812
Fermentation	**5:**864
Hormones	**6:**1050
Human Genome Project	**6:**1060
Lipids	**6:**1191
Metabolism	**7:**1255
Nucleic acid	**7:**1387
Osmosis	**7:**1436
Photosynthesis	**8:**1505
Proteins	**8:**1586
Respiration	**9:**1672
Vitamin	**10:**1981
Yeast	**10:**2043

Biology

Adaptation	**1:**26
Algae	**1:**91
Amino acid	**1:**130
Amoeba	**1:**131
Amphibians	**1:**134
Anatomy	**1:**138
Animal	**1:**145
Antibody and antigen	**1:**159
Arachnids	**1:**168
Arthropods	**1:**183
Bacteria	**2:**253
Behavior	**2:**270
Biochemistry	**2:**279
Biodegradable	**2:**280
Biodiversity	**2:**281
Biological warfare	**2:**287
Biology	**2:**290
Biome	**2:**293
Biophysics	**2:**302
Biosphere	**2:**304
Biotechnology	**2:**309
Birds	**2:**312
Birth	**2:**315
Birth defects	**2:**319
Blood	**2:**326
Botany	**2:**334
Brain	**2:**337
Butterflies	**2:**364
Canines	**2:**382
Carbohydrate	**2:**387
Carcinogen	**2:**406
Cell	**3:**428
Cellulose	**3:**442
Cetaceans	**3:**448
Cholesterol	**3:**469
Chromosome	**3:**472
Circulatory system	**3:**480
Clone and cloning	**3:**484
Cockroaches	**3:**505
Coelacanth	**3:**508
Contraception	**3:**562
Coral	**3:**566
Crustaceans	**3:**590
Cryobiology	**3:**593
Digestive system	**4:**653
Dinosaur	**4:**658
Disease	**4:**669
Ear	**4:**693
Embryo and embryonic development	**4:**785
Endocrine system	**5:**796
Enzyme	**5:**812
Eutrophication	**5:**828
Evolution	**5:**832
Excretory system	**5:**839
Eye	**5:**848
Felines	**5:**855
Fermentation	**5:**864
Fertilization	**5:**867
Fish	**5:**875
Flower	**5:**878
Forestry	**5:**901
Forests	**5:**907
Fungi	**5:**930
Genetic disorders	**5:**966
Genetic engineering	**5:**973
Genetics	**5:**980
Heart	**6:**1037
Hibernation	**6:**1046
Hormones	**6:**1050
Horticulture	**6:**1053
Human Genome Project	**6:**1060
Human evolution	**6:**1054
Immune system	**6:**1082
Indicator species	**6:**1090
Insects	**6:**1103
Integumentary system	**6:**1109
Invertebrates	**6:**1133
Kangaroos and wallabies	**6:**1153
Leaf	**6:**1172
Lipids	**6:**1191
Lymphatic system	**6:**1198
Mammals	**6:**1222

Entries by Scientific Field

Entries by Scientific Field

Mendelian laws of inheritance	7:1246	Vaccine	10:1957
Metabolism	7:1255	Vertebrates	10:1967
Metamorphosis	7:1259	Virus	10:1974
Migration (animals)	7:1271	Vitamin	10:1981
Molecular biology	7:1283	Wetlands	10:2024
Mollusks	7:1288	Yeast	10:2043
Muscular system	7:1309		
Mutation	7:1314		
Nervous system	7:1333		
Nucleic acid	7:1387		
Osmosis	7:1436		
Parasites	8:1467		
Photosynthesis	8:1505		
Phototropism	8:1508		
Physiology	8:1516		
Plague	8:1518		
Plankton	8:1520		
Plant	8:1522		
Primates	8:1571		
Proteins	8:1586		
Protozoa	8:1590		
Puberty	8:1599		
Rain forest	8:1641		
Reproduction	9:1664		
Reproductive system	9:1667		
Reptiles	9:1670		
Respiration	9:1672		
Respiratory system	9:1677		
Rh factor	9:1683		
Seed	9:1729		
Sexually transmitted diseases	9:1735		
Skeletal system	9:1739		
Smell	9:1750		
Snakes	9:1752		
Speech	9:1796		
Sponges	9:1799		
Taste	10:1861		
Touch	10:1903		
Tree	10:1927		
Tumor	10:1934		

Biomedical engineering

Electrocardiogram	4:751
Radiology	8:1637

Biotechnology

Biotechnology	2:309
Brewing	2:352
Fermentation	5:864
Vaccine	10:1957

Botany

Botany	2:334
Cellulose	3:442
Cocaine	3:501
Cotton	3:577
Flower	5:878
Forestry	5:901
Forests	5:907
Horticulture	6:1053
Leaf	6:1172
Marijuana	6:1224
Photosynthesis	8:1505
Phototropism	8:1508
Plant	8:1522
Seed	9:1729
Tree	10:1927

Cartography

Cartography	2:410
Geologic map	5:986

Entries by Scientific Field

Cellular biology

Amino acid	1:130
Carbohydrate	2:387
Cell	3:428
Cholesterol	3:469
Chromosome	3:472
Genetics	5:980
Lipids	6:1191
Osmosis	7:1436
Proteins	8:1586

Chemistry

Acids and bases	1:14
Actinides	1:23
Aerosols	1:43
Agent Orange	1:54
Agrochemical	1:65
Alchemy	1:82
Alcohols	1:88
Alkali metals	1:99
Alkaline earth metals	1:102
Aluminum family	1:122
Atom	2:226
Atomic mass	2:229
Atomic theory	2:232
Biochemistry	2:279
Carbon dioxide	2:393
Carbon family	2:395
Carbon monoxide	2:403
Catalyst and catalysis	2:413
Chemical bond	3:453
Chemical w\arfare	3:457
Chemistry	3:463
Colloid	3:515
Combustion	3:522
Composite materials	3:536
Compound, chemical	3:541
Crystal	3:601
Cyclamate	3:608
DDT (dichlorodiphenyl-trichloroethane)	4:619
Diffusion	4:651
Dioxin	4:667
Distillation	4:675
Dyes and pigments	4:686
Electrolysis	4:755
Element, chemical	4:774
Enzyme	5:812
Equation, chemical	5:815
Equilibrium, chemical	5:817
Explosives	5:843
Fermentation	5:864
Filtration	5:872
Formula, chemical	5:914
Halogens	6:1030
Hormones	6:1050
Hydrogen	6:1068
Industrial minerals	6:1092
Ionization	6:1135
Isotope	6:1141
Lanthanides	6:1163
Lipids	6:1191
Metabolism	7:1255
Mole (measurement)	7:1282
Molecule	7:1285
Nitrogen family	7:1344
Noble gases	7:1349
Nucleic acid	7:1387
Osmosis	7:1436
Oxidation-reduction reaction	7:1439
Oxygen family	7:1442
Ozone	7:1450
Periodic table	8:1486
pH	8:1495
Photochemistry	8:1498
Photosynthesis	8:1505
Plastics	8:1532
Poisons and toxins	8:1542
Polymer	8:1563
Proteins	8:1586
Qualitative analysis	8:1603
Quantitative analysis	8:1604

Entries by Scientific Field

Reaction, chemical	**9:**1647
Respiration	**9:**1672
Soaps and detergents	**9:**1756
Solution	**9:**1767
Transition elements	**10:**1913
Vitamin	**10:**1981
Yeast	**10:**2043

Civil engineering

Bridges	**2:**354
Canal	**2:**376
Dam	**4:**611
Lock	**6:**1192

Climatology

Global climate	**5:**1006
Ice ages	**6:**1075
Seasons	**9:**1726

Communications/ Graphic arts

Antenna	**1:**153
CAD/CAM	**2:**369
Cellular/digital technology	**3:**439
Compact disc	**3:**531
Computer software	**3:**549
DVD technology	**4:**684
Hologram and holography	**6:**1048
Internet	**6:**1123
Magnetic recording/ audiocassette	**6:**1209
Microwave communication	**7:**1268
Petroglyphs and pictographs	**8:**1491
Photocopying	**8:**1499
Radio	**8:**1626
Satellite	**9:**1707
Telegraph	**10:**1863
Telephone	**10:**1866
Television	**10:**1875
Video recording	**10:**1968

Computer science

Artificial intelligence	**1:**188
Automation	**2:**242
CAD/CAM	**2:**369
Calculator	**2:**370
Cellular/digital technology	**3:**439
Compact disc	**3:**531
Computer, analog	**3:**546
Computer, digital	**3:**547
Computer software	**3:**549
Internet	**6:**1123
Mass production	**7:**1236
Robotics	**9:**1690
Virtual reality	**10:**1969

Cosmology

Astrophysics	**1:**207
Big Bang theory	**2:**273
Cosmology	**3:**574
Galaxy	**5:**941
Space	**9:**1776

Cryogenics

Cryobiology	**3:**593
Cryogenics	**3:**595

Dentistry

Dentistry	**4:**626
Fluoridation	**5:**889

Ecology/Environmental science

Acid rain	**1:**9
Alternative energy sources	**1:**111
Biodegradable	**2:**280
Biodiversity	**2:**281

Entries by Scientific Field

Bioenergy	**2:**284
Biome	**2:**293
Biosphere	**2:**304
Carbon cycle	**2:**389
Composting	**3:**539
DDT (dichlorodiphenyl-trichloroethane)	**4:**619
Desert	**4:**634
Dioxin	**4:**667
Drift net	**4:**680
Drought	**4:**682
Ecology	**4:**725
Ecosystem	**4:**728
Endangered species	**5:**793
Environmental ethics	**5:**807
Erosion	**5:**820
Eutrophication	**5:**828
Food web and food chain	**5:**894
Forestry	**5:**901
Forests	**5:**907
Gaia hypothesis	**5:**935
Greenhouse effect	**5:**1016
Hydrologic cycle	**6:**1071
Indicator species	**6:**1090
Nitrogen cycle	**7:**1342
Oil spills	**7:**1422
Organic farming	**7:**1431
Paleoecology	**8:**1457
Pollution	**8:**1549
Pollution control	**8:**1558
Rain forest	**8:**1641
Recycling	**9:**1650
Succession	**10:**1837
Waste management	**10:**2003
Wetlands	**10:**2024

Electrical engineering

Antenna	**1:**153
Battery	**2:**268
Cathode	**3:**415
Cathode-ray tube	**3:**417
Cell, electrochemical	**3:**436
Compact disc	**3:**531
Diode	**4:**665
Electric arc	**4:**734
Electric current	**4:**737
Electricity	**4:**741
Electric motor	**4:**747
Electrocardiogram	**4:**751
Electromagnetic field	**4:**758
Electromagnetic induction	**4:**760
Electromagnetism	**4:**766
Electronics	**4:**773
Fluorescent light	**5:**886
Generator	**5:**962
Incandescent light	**6:**1087
Integrated circuit	**6:**1106
LED (light-emitting diode)	**6:** 1176
Magnetic recording/audiocassette	**6:**1209
Radar	**8:**1613
Radio	**8:**1626
Superconductor	**10:**1849
Telegraph	**10:**1863
Telephone	**10:**1866
Television	**10:**1875
Transformer	**10:**1908
Transistor	**10:**1910
Ultrasonics	**10:**1941
Video recording	**10:**1968

Electronics

Antenna	**1:**153
Battery	**2:**268
Cathode	**3:**415
Cathode-ray tube	**3:**417
Cell, electrochemical	**3:**436
Compact disc	**3:**531
Diode	**4:**665
Electric arc	**4:**734
Electric current	**4:**737
Electricity	**4:**741
Electric motor	**4:**747

Entries by Scientific Field

Electromagnetic field	**4:**758
Electromagnetic induction	**4:**760
Electronics	**4:**773
Generator	**5:**962
Integrated circuit	**6:**1106
LED (light-emitting diode)	**6:**1176
Magnetic recording/ audiocassette	**6:**1209
Radar	**8:**1613
Radio	**8:**1626
Superconductor	**10:**1849
Telephone	**10:**1866
Television	**10:**1875
Transformer	**10:**1908
Transistor	**10:**1910
Ultrasonics	**10:**1941
Video recording	**10:**1968

Embryology

Embryo and embryonic development	**4:**785
Fertilization	**5:**867
Reproduction	**9:**1664
Reproductive system	**9:**1667

Engineering

Aerodynamics	**1:**39
Aircraft	**1:**74
Antenna	**1:**153
Automation	**2:**242
Automobile	**2:**245
Balloon	**1:**261
Battery	**2:**268
Bridges	**2:**354
Canal	**2:**376
Cathode	**3:**415
Cathode-ray tube	**3:**417
Cell, electrochemical	**3:**436
Compact disc	**3:**531
Dam	**4:**611
Diesel engine	**4:**646
Diode	**4:**665
Electric arc	**4:**734
Electric current	**4:**737
Electric motor	**4:**747
Electricity	**4:**741
Electrocardiogram	**4:**751
Electromagnetic field	**4:**758
Electromagnetic induction	**4:**760
Electromagnetism	**4:**766
Electronics	**4:**773
Engineering	**5:**805
Fluorescent light	**5:**886
Generator	**5:**962
Incandescent light	**6:**1087
Integrated circuit	**6:**1106
Internal-combustion engine	**6:**1117
Jet engine	**6:**1143
LED (light-emitting diode)	**6:** 1176
Lock	**6:**1192
Machines, simple	**6:**1203
Magnetic recording/ audiocassette	**6:**1209
Mass production	**7:**1236
Radar	**8:**1613
Radio	**8:**1626
Steam engine	**9:**1817
Submarine	**10:**1834
Superconductor	**10:**1849
Telegraph	**10:**1863
Telephone	**10:**1866
Television	**10:**1875
Transformer	**10:**1908
Transistor	**10:**1910
Ultrasonics	**10:**1941
Video recording	**10:**1968

Entomology

Arachnids	**1:**168
Arthropods	**1:**183

Entries by Scientific Field

Butterflies	2:364
Cockroaches	3:505
Insects	6:1103
Invertebrates	6:1133
Metamorphosis	7:1259

Epidemiology

Biological warfare	2:287
Disease	4:669
Ebola virus	4:717
Plague	8:1518
Poliomyelitis	8:1546
Sexually transmitted diseases	9:1735
Vaccine	10:1957

Evolutionary biology

Adaptation	1:26
Evolution	5:832
Human evolution	6:1054
Mendelian laws of inheritance	7:1246

Food science

Brewing	2:352
Cyclamate	3:608
Food preservation	5:890
Nutrition	7:1399

Forensic science

Forensic science	5:898

Forestry

Forestry	5:901
Forests	5:907
Rain forest	8:1641
Tree	10:1927

General science

Alchemy	1:82
Chaos theory	3:451
Metric system	7:1265
Scientific method	9:1722
Units and standards	10:1948

Genetic engineering

Biological warfare	2:287
Biotechnology	2:309
Clone and cloning	3:484
Genetic engineering	5:973

Genetics

Biotechnology	2:309
Birth defects	2:319
Cancer	2:379
Carcinogen	2:406
Chromosome	3:472
Clone and cloning	3:484
Genetic disorders	5:966
Genetic engineering	5:973
Genetics	5:980
Human Genome Project	6:1060
Mendelian laws of inheritance	7:1246
Mutation	7:1314
Nucleic acid	7:1387

Geochemistry

Coal	3:492
Earth (planet)	4:698
Earth science	4:707
Earth's interior	4:708
Glacier	5:1000
Minerals	7:1273
Rocks	9:1701
Soil	9:1758

Entries by Scientific Field

Geography

Africa	1:49
Antarctica	1:147
Asia	1:194
Australia	2:238
Biome	2:293
Cartography	2:410
Coast and beach	3:498
Desert	4:634
Europe	5:823
Geologic map	5:986
Island	6:1137
Lake	6:1159
Mountain	7:1301
North America	7:1352
River	9:1685
South America	9:1772

Geology

Catastrophism	3:415
Cave	3:420
Coal	3:492
Coast and beach	3:498
Continental margin	3:560
Dating techniques	4:616
Desert	4:634
Earthquake	4:702
Earth science	4:707
Earth's interior	4:708
Erosion	5:820
Fault	5:855
Geologic map	5:986
Geologic time	5:990
Geology	5:993
Glacier	5:1000
Hydrologic cycle	6:1071
Ice ages	6:1075
Iceberg	6:1078
Industrial minerals	6:1092
Island	6:1137
Lake	6:1159
Minerals	7:1273
Mining	7:1278
Mountain	7:1301
Natural gas	7:1319
Oil drilling	7:1418
Oil spills	7:1422
Petroleum	8:1492
Plate tectonics	8:1534
River	9:1685
Rocks	9:1701
Soil	9:1758
Uniformitarianism	10:1946
Volcano	10:1992
Water	10:2010

Geophysics

Earth (planet)	4:698
Earth science	4:707
Fault	5:855
Plate tectonics	8:1534

Gerontology

Aging and death	1:59
Alzheimer's disease	1:126
Arthritis	1:181
Dementia	4:622
Gerontology	5:999

Gynecology

Contraception	3:562
Fertilization	5:867
Gynecology	5:1022
Puberty	8:1599
Reproduction	9:1664

Health/Medicine

Acetylsalicylic acid	1:6
Addiction	1:32
Attention-deficit hyperactivity disorder (ADHD)	2:237

Entries by Scientific Field

Depression	**4:**630	Hallucinogens	**6:**1027
AIDS (acquired immunod-eficiency syndrome)	**1:**70	Immune system	**6:**1082
		Legionnaire's disease	**6:**1179
Alcoholism	**1:**85	Lipids	**6:**1191
Allergy	**1:**106	Malnutrition	**6:**1216
Alternative medicine	**1:**118	Marijuana	**6:**1224
Alzheimer's disease	**1:**126	Multiple personality disorder	**7:**1305
Amino acid	**1:**130		
Anesthesia	**1:**142	Nuclear medicine	**7:**1372
Antibiotics	**1:**155	Nutrition	**7:**1399
Antiseptics	**1:**164	Obsession	**7:**1405
Arthritis	**1:**181	Orthopedics	**7:**1434
Asthma	**1:**204	Parasites	**8:**1467
Attention-deficit hyperactivity disorder (ADHD)	**2:**237	Phobia	**8:**1497
		Physical therapy	**8:**1511
Birth defects	**2:**319	Plague	**8:**1518
Blood supply	**2:**330	Plastic surgery	**8:**1527
Burn	**2:**361	Poliomyelitis	**8:**1546
Carcinogen	**2:**406	Prosthetics	**8:**1579
Carpal tunnel syndrome	**2:**408	Protease inhibitor	**8:**1583
Cholesterol	**3:**469	Psychiatry	**8:**1592
Cigarette smoke	**3:**476	Psychology	**8:**1594
Cocaine	**3:**501	Psychosis	**8:**1596
Contraception	**3:**562	Puberty	**8:**1599
Dementia	**4:**622	Radial keratotomy	**8:**1615
Dentistry	**4:**626	Radiology	**8:**1637
Depression	**4:**630	Rh factor	**9:**1683
Diabetes mellitus	**4:**638	Schizophrenia	**9:**1716
Diagnosis	**4:**640	Sexually transmitted diseases	**9:**1735
Dialysis	**4:**644		
Disease	**4:**669	Sleep and sleep disorders	**9:**1745
Dyslexia	**4:**690	Stress	**9:**1826
Eating disorders	**4:**711	Sudden infant death syndrome (SIDS)	**10:**1840
Ebola virus	**4:**717		
Electrocardiogram	**4:**751	Surgery	**10:**1855
Fluoridation	**5:**889	Tranquilizers	**10:**1905
Food preservation	**5:**890	Transplant, surgical	**10:**1923
Genetic disorders	**5:**966	Tumor	**10:**1934
Genetic engineering	**5:**973	Vaccine	**10:**1957
Genetics	**5:**980	Virus	**10:**1974
Gerontology	**5:**999	Vitamin	**10:**1981
Gynecology	**5:**1022	Vivisection	**10:**1989

Entries by Scientific Field

Horticulture

Horticulture	**6:**1053
Plant	**8:**1522
Seed	**9:**1729
Tree	**10:**1927

Immunology

Allergy	**1:**106
Antibiotics	**1:**155
Antibody and antigen	**1:**159
Immune system	**6:**1082
Vaccine	**10:**1957

Marine biology

Algae	**1:**91
Amphibians	**1:**134
Cetaceans	**3:**448
Coral	**3:**566
Crustaceans	**3:**590
Endangered species	**5:**793
Fish	**5:**875
Mammals	**6:**1222
Mollusks	**7:**1288
Ocean zones	**7:**1414
Plankton	**8:**1520
Sponges	**9:**1799
Vertebrates	**10:**1967

Materials science

Abrasives	**1:**2
Adhesives	**1:**37
Aerosols	**1:**43
Alcohols	**1:**88
Alkaline earth metals	**1:**102
Alloy	**1:**110
Aluminum family	**1:**122
Artificial fibers	**1:**186
Asbestos	**1:**191
Biodegradable	**2:**280
Carbon family	**2:**395
Ceramic	**3:**447
Composite materials	**3:**536
Dyes and pigments	**4:**686
Electrical conductivity	**4:**731
Electrolysis	**4:**755
Expansion, thermal	**5:**842
Fiber optics	**5:**870
Glass	**5:**1004
Halogens	**6:**1030
Hand tools	**6:**1036
Hydrogen	**6:**1068
Industrial minerals	**6:**1092
Minerals	**7:**1273
Nitrogen family	**7:**1344
Oxygen family	**7:**1442
Plastics	**8:**1532
Polymer	**8:**1563
Soaps and detergents	**9:**1756
Superconductor	**10:**1849
Transition elements	**10:**1913

Mathematics

Abacus	**1:**1
Algebra	**1:**97
Arithmetic	**1:**177
Boolean algebra	**2:**333
Calculus	**2:**371
Chaos theory	**3:**451
Circle	**3:**478
Complex numbers	**3:**534
Correlation	**3:**569
Fractal	**5:**921
Fraction, common	**5:**923
Function	**5:**927
Game theory	**5:**945
Geometry	**5:**995
Graphs and graphing	**5:**1009
Imaginary number	**6:**1081
Logarithm	**6:**1195
Mathematics	**7:**1241

Entries by Scientific Field

Multiplication 7:1307
Natural numbers 7:1321
Number theory 7:1393
Numeration systems 7:1395
Polygon 8:1562
Probability theory 8:1575
Proof (mathematics) 8:1578
Pythagorean theorem 8:1601
Set theory 9:1733
Statistics 9:1810
Symbolic logic 10:1859
Topology 10:1897
Trigonometry 10:1931
Zero 10:2047

Metallurgy

Alkali metals 1:99
Alkaline earth metals 1:102
Alloy 1:110
Aluminum family 1:122
Carbon family 2:395
Composite materials 3:536
Industrial minerals 6:1092
Minerals 7:1273
Mining 7:1278
Precious metals 8:1566
Transition elements 10:1913

Meteorology

Air masses and fronts 1:80
Atmosphere, composition and structure 2:211
Atmosphere observation 2:215
Atmospheric circulation 2:218
Atmospheric optical effects 2:221
Atmospheric pressure 2:225
Barometer 2:265
Clouds 3:490
Cyclone and anticyclone 3:608
Drought 4:682
El Niño 4:782
Global climate 5:1006
Monsoon 7:1291
Ozone 7:1450
Storm surge 9:1823
Thunderstorm 10:1887
Tornado 10:1900
Weather 10:2017
Weather forecasting 10:2020
Wind 10:2028

Microbiology

Algae 1:91
Amoeba 1:131
Antiseptics 1:164
Bacteria 2:253
Biodegradable 2:280
Biological warfare 2:287
Composting 3:539
Parasites 8:1467
Plankton 8:1520
Protozoa 8:1590
Yeast 10:2043

Mineralogy

Abrasives 1:2
Ceramic 3:447
Industrial minerals 6:1092
Minerals 7:1273
Mining 7:1278

Molecular biology

Amino acid 1:130
Antibody and antigen 1:159
Biochemistry 2:279
Birth defects 2:319
Chromosome 3:472
Clone and cloning 3:484
Enzyme 5:812
Genetic disorders 5:966

Entries by Scientific Field

Genetic engineering **5:**973
Genetics **5:**980
Hormones **6:**1050
Human Genome Project **6:**1060
Lipids **6:**1191
Molecular biology **7:**1283
Mutation **7:**1314
Nucleic acid **7:**1387
Proteins **8:**1586

Mycology

Brewing **2:**352
Fermentation **5:**864
Fungi **5:**930
Yeast **10:**2043

Nutrition

Diabetes mellitus **4:**638
Eating disorders **4:**711
Food web and food chain **5:**894
Malnutrition **6:**1216
Nutrition **7:**1399
Vitamin **10:**1981

Obstetrics

Birth **2:**315
Birth defects **2:**319
Embryo and embryonic development **4:**785

Oceanography

Continental margin **3:**560
Currents, ocean **3:**604
Ocean **7:**1407
Oceanography **7:**1411
Ocean zones **7:**1414
Tides **10:**1890

Oncology

Cancer **2:**379
Disease **4:**669
Tumor **10:**1934

Ophthalmology

Eye **5:**848
Lens **6:**1184
Radial keratotomy **8:**1615

Optics

Atmospheric optical effects **2:**221
Compact disc **3:**531
Diffraction **4:**648
Eye **5:**848
Fiber optics **5:**870
Hologram and holography **6:**1048
Laser **6:**1166
LED (light-emitting diode) **6:**1176
Lens **6:**1184
Light **6:**1185
Luminescence **6:**1196
Photochemistry **8:**1498
Photocopying **8:**1499
Telescope **10:**1869
Television **10:**1875
Video recording **10:**1968

Organic chemistry

Carbon family **2:**395
Coal **3:**492
Cyclamate **3:**608
Dioxin **4:**667
Fermentation **5:**864
Hydrogen **6:**1068
Hydrologic cycle **6:**1071
Lipids **6:**1191

Entries by Scientific Field

Natural gas 7:1319
Nitrogen cycle 7:1342
Nitrogen family 7:1344
Oil spills 7:1422
Organic chemistry 7:1428
Oxygen family 7:1442
Ozone 7:1450
Petroleum 8:1492
Vitamin 10:1981

Orthopedics

Arthritis 1:181
Orthopedics 7:1434
Prosthetics 8:1579
Skeletal system 9:1739

Paleontology

Dating techniques 4:616
Dinosaur 4:658
Evolution 5:832
Fossil and fossilization 5:917
Human evolution 6:1054
Paleoecology 8:1457
Paleontology 8:1459

Parasitology

Amoeba 1:131
Disease 4:669
Fungi 5:930
Parasites 8:1467

Pathology

AIDS (acquired immunode-
 ficiency syndrome) 1:70
Alzheimer's disease 1:126
Arthritis 1:181
Asthma 1:204
Attention-deficit hyperactivity
 disorder (ADHD) 2:237
Bacteria 2:253
Biological warfare 2:287
Cancer 2:379
Dementia 4:622
Diabetes mellitus 4:638
Diagnosis 4:640
Dioxin 4:667
Disease 4:669
Ebola virus 4:717
Genetic disorders 5:966
Malnutrition 6:1216
Orthopedics 7:1434
Parasites 8:1467
Plague 8:1518
Poliomyelitis 8:1546
Sexually transmitted
 diseases 9:1735
Tumor 10:1934
Vaccine 10:1957
Virus 10:1974

Pharmacology

Acetylsalicylic acid 1:6
Antibiotics 1:155
Antiseptics 1:164
Cocaine 3:501
Hallucinogens 6:1027
Marijuana 6:1224
Poisons and toxins 8:1542
Tranquilizers 10:1905

Physics

Acceleration 1:4
Acoustics 1:17
Aerodynamics 1:39
Antiparticle 1:163
Astrophysics 1:207
Atom 2:226
Atomic mass 2:229
Atomic theory 2:232
Ballistics 2:260

Entries by Scientific Field

Battery	2:268	Gases, properties of	5:959
Biophysics	2:302	Generator	5:962
Buoyancy	2:360	Gravity and gravitation	5:1012
Calorie	2:375	Gyroscope	5:1024
Cathode	3:415	Half-life	6:1027
Cathode-ray tube	3:417	Heat	6:1043
Celestial mechanics	3:423	Hologram and holography	6:1048
Cell, electrochemical	3:436	Incandescent light	6:1087
Chaos theory	3:451	Integrated circuit	6:1106
Color	3:518	Interference	6:1112
Combustion	3:522	Interferometry	6:1114
Conservation laws	3:554	Ionization	6:1135
Coulomb	3:579	Isotope	6:1141
Cryogenics	3:595	Laser	6:1166
Dating techniques	4:616	Laws of motion	6:1169
Density	4:624	LED (light-emitting diode)	6:1176
Diffraction	4:648	Lens	6:1184
Diode	4:665	Light	6:1185
Doppler effect	4:677	Luminescence	6:1196
Echolocation	4:720	Magnetic recording/	
Elasticity	4:730	audiocassette	6:1209
Electrical conductivity	4:731	Magnetism	6:1212
Electric arc	4:734	Mass	7:1235
Electric current	4:737	Mass spectrometry	7:1239
Electricity	4:741	Matter, states of	7:1243
Electric motor	4:747	Microwave communication	7:1268
Electrolysis	4:755	Molecule	7:1285
Electromagnetic field	4:758	Momentum	7:1290
Electromagnetic induction	4:760	Nuclear fission	7:1361
Electromagnetic spectrum	4:763	Nuclear fusion	7:1366
Electromagnetism	4:766	Nuclear medicine	7:1372
Electron	4:768	Nuclear power	7:1374
Electronics	4:773	Nuclear weapons	7:1381
Energy	5:801	Particle accelerators	8:1475
Evaporation	5:831	Periodic function	8:1485
Expansion, thermal	5:842	Photochemistry	8:1498
Fiber optics	5:870	Photoelectric effect	8:1502
Fluid dynamics	5:882	Physics	8:1513
Fluorescent light	5:886	Pressure	8:1570
Frequency	5:925	Quantum mechanics	8:1607
Friction	5:926	Radar	8:1613
Gases, liquefaction of	5:955	Radiation	8:1619

Entries by Scientific Field

Radiation exposure	**8:**1621
Radio	**8:**1626
Radioactive tracers	**8:**1629
Radioactivity	**8:**1630
Radiology	**8:**1637
Relativity, theory of	**9:**1659
Sonar	**9:**1770
Spectroscopy	**9:**1792
Spectrum	**9:**1794
Subatomic particles	**10:**1829
Superconductor	**10:**1849
Telegraph	**10:**1863
Telephone	**10:**1866
Television	**10:**1875
Temperature	**10:**1879
Thermal expansion	**5:**842
Thermodynamics	**10:**1885
Time	**10:**1894
Transformer	**10:**1908
Transistor	**10:**1910
Tunneling	**10:**1937
Ultrasonics	**10:**1941
Vacuum	**10:**1960
Vacuum tube	**10:**1961
Video recording	**10:**1968
Virtual reality	**10:**1969
Volume	**10:**1999
Wave motion	**10:**2014
X ray	**10:**2033

Primatology

Animal	**1:**145
Endangered species	**5:**793
Mammals	**6:**1222
Primates	**8:**1571
Vertebrates	**10:**1967

Psychiatry/Psychology

Addiction	**1:**32
Alcoholism	**1:**85
Attention-deficit hyperactivity disorder (ADHD)	**2:**237
Behavior	**2:**270
Cognition	**3:**511
Depression	**4:**630
Eating disorders	**4:**711
Multiple personality disorder	**7:**1305
Obsession	**7:**1405
Perception	**8:**1482
Phobia	**8:**1497
Psychiatry	**8:**1592
Psychology	**8:**1594
Psychosis	**8:**1596
Reinforcement, positive and negative	**9:**1657
Savant	**9:**1712
Schizophrenia	**9:**1716
Sleep and sleep disorders	**9:**1745
Stress	**9:**1826

Radiology

Nuclear medicine	**7:**1372
Radioactive tracers	**8:**1629
Radiology	**8:**1637
Ultrasonics	**10:**1941
X ray	**10:**2033

Robotics

Automation	**2:**242
Mass production	**7:**1236
Robotics	**9:**1690

Seismology

Earthquake	**4:**702
Volcano	**10:**1992

Sociology

Adaptation	**1:**26
Aging and death	**1:**59

Entries by Scientific Field

Alcoholism	1:85	Dyes and pigments	4:686
Behavior	2:270	Fiber optics	5:870
Gerontology	5:999	Fluorescent light	5:886
Migration (animals)	7:1271	Food preservation	5:890
		Forensic science	5:898
		Generator	5:962

Technology

		Glass	5:1004
Abrasives	1:2	Hand tools	6:1036
Adhesives	1:37	Hologram and holography	6:1048
Aerosols	1:43	Incandescent light	6:1087
Aircraft	1:74	Industrial Revolution	6:1097
Alloy	1:110	Integrated circuit	6:1106
Alternative energy sources	1:111	Internal-combustion engine	6:1117
Antenna	1:153	Internet	6:1123
Artificial fibers	1:186	Jet engine	6:1143
Artificial intelligence	1:188	Laser	6:1166
Asbestos	1:191	LED (light-emitting diode)	6:1176
Automation	2:242	Lens	6:1184
Automobile	2:245	Lock	6:1192
Balloon	1:261	Machines, simple	6:1203
Battery	2:268	Magnetic recording/	
Biotechnology	2:309	audiocassette	6:1209
Brewing	2:352	Mass production	7:1236
Bridges	2:354	Mass spectrometry	7:1239
CAD/CAM	2:369	Microwave communication	7:1268
Calculator	2:370	Paper	8:1462
Canal	2:376	Photocopying	8:1499
Cathode	3:415	Plastics	8:1532
Cathode-ray tube	3:417	Polymer	8:1563
Cell, electrochemical	3:436	Prosthetics	8:1579
Cellular/digital technology	3:439	Radar	8:1613
Centrifuge	3:445	Radio	8:1626
Ceramic	3:447	Robotics	9:1690
Compact disc	3:531	Rockets and missiles	9:1693
Computer, analog	3:546	Soaps and detergents	9:1756
Computer, digital	3:547	Sonar	9:1770
Computer software	3:549	Space station, international	9:1788
Cybernetics	3:605	Steam engine	9:1817
Dam	4:611	Submarine	10:1834
Diesel engine	4:646	Superconductor	10:1849
Diode	4:665	Telegraph	10:1863
DVD technology	4:684	Telephone	10:1866

Entries by Scientific Field

Television	**10:**1875
Transformer	**10:**1908
Transistor	**10:**1910
Vacuum tube	**10:**1961
Video recording	**10:**1968
Virtual reality	**10:**1969

Virology

AIDS (acquired immuno-deficiency syndrome)	**1:**70
Disease	**4:**669
Ebola virus	**4:**717
Plague	**8:**1518
Poliomyelitis	**8:**1546
Sexually transmitted diseases	**9:**1735
Vaccine	**10:**1957
Virus	**10:**1974

Weaponry

Ballistics	**2:**260
Biological warfare	**2:**287
Chemical warfare	**3:**457
Forensic science	**5:**898
Nuclear weapons	**7:**1381
Radar	**8:**1613
Rockets and missiles	**9:**1693

Wildlife conservation

Biodiversity	**2:**281
Biome	**2:**293
Biosphere	**2:**304
Drift net	**4:**680
Ecology	**4:**725
Ecosystem	**4:**728
Endangered species	**5:**793
Forestry	**5:**901
Gaia hypothesis	**5:**935
Wetlands	**10:**2024

Zoology

Amphibians	**1:**134
Animal	**1:**145
Arachnids	**1:**168
Arthropods	**1:**183
Behavior	**2:**270
Birds	**2:**312
Butterflies	**2:**364
Canines	**2:**382
Cetaceans	**3:**448
Cockroaches	**3:**505
Coelacanth	**3:**508
Coral	**3:**566
Crustaceans	**3:**590
Dinosaur	**4:**658
Echolocation	**4:**720
Endangered species	**5:**793
Felines	**5:**855
Fish	**5:**875
Hibernation	**6:**1046
Indicator species	**6:**1090
Insects	**6:**1103
Invertebrates	**6:**1133
Kangaroos and wallabies	**6:**1153
Mammals	**6:**1222
Metamorphosis	**7:**1259
Migration (animals)	**7:**1271
Mollusks	**7:**1288
Plankton	**8:**1520
Primates	**8:**1571
Reptiles	**9:**1670
Snakes	**9:**1752
Sponges	**9:**1799
Vertebrates	**10:**1967

Mass

One common method of defining mass is to say that it is the quantity of matter an object possesses. For example, a small rock has a fixed, unchanging quantity of matter. If you were to take that rock to the Moon, to Mars, or to any other part of the universe, it would have the same quantity of matter—the same mass—as it has on Earth.

Mass is sometimes confused with weight. Weight is defined as the gravitational attraction on an object by some body, such as Earth or the Moon. The rock described above would have a greater weight on Earth than on the Moon because Earth exerts a greater gravitational attraction on bodies than does the Moon.

Mass and the second law

A more precise definition of mass can be obtained from Newton's second law of motion. According to that law—and assuming that the object in question is free to move horizontally without friction—if a constant force is applied to an object, that object will gain speed. For example, if you hit a ball with a hammer (the constant force), the ball goes from a zero velocity (when it is at rest) to some speed as it rolls across the ground. Mathematically, the second law can be written as $F = m \cdot a$, where F is the force used to move an object, m is the mass of the object, and a is the acceleration, or increase in speed of the object.

Newton's second law says that the amount of speed gained by an object when struck by a force depends on the quantity of matter in the object. Suppose that you strike a bowling ball and a golf ball with the same force. The golf ball gains a great deal more speed than does the bowling ball because it takes a greater force to get the bowling ball moving than it does to get the golf ball moving.

This fact provides another way of defining mass. Mass is the increase in speed of an object provided by some given force. Or, one can solve the equation above for m, the mass of an object, to get m = F ÷ a. A kilogram, for example, can be defined as the mass that increases its speed at the rate of one meter per second when it is struck by a force of one newton.

Units of mass

In the SI system of measurement (the International System of Units), the fundamental unit of mass is the kilogram. A smaller unit, the gram, is also used widely in many measurements. In the English system, the unit of mass is the slug. A slug is equal to 14.6 kilograms.

Scientists and nonscientists alike commonly convert measurements between kilogram and pounds, not kilograms and slugs. Technically, though, a kilogram/pound conversion is not correct since kilogram is a measure of mass and pound a measure of weight. However, such measurements and such conversions almost always involve observations made on Earth's surface where there is a constant ratio between mass and weight.

[*See also* **Acceleration; Density; Force; Laws of motion; Matter, states of**]

Mass production

Mass production is the manufacture of goods in large quantities using standardized designs so the goods are all the same. Assembly-line techniques are usually used. An assembly line is a system in which a product is manufactured in a step-by-step process as it moves continuously past an arrangement of workers and machines. This system is one of the most powerful productivity concepts in history. It was largely responsible for the emergence and expansion of the industrialized, consumer-based system we have today.

While various mass production techniques were practiced in ancient times, the English were probably the first to use water-powered and steam-powered machinery in industrial production during the Industrial Revolution that began in the mid-1700s. But it is generally agreed that mod-

> **Words to Know**
>
> **Assembly line:** A sequence of workers, machines, and parts down which an incomplete product passes, each worker performing a procedure, until the product is assembled.
>
> **Interchangeability:** Parts that are so similar that they can be switched between different machines or products and the machines or products will still work.

ern mass production techniques came into widespread use through the inventiveness of Americans. As a matter of fact, modern mass production has been called the "American System."

Famous American contributors to mass production

The early successes of the American System are often attributed to Eli Whitney. He adapted mass production techniques and the interchangeability of parts to the manufacture of muskets (a type of gun) for the U.S. government in the 1790s.

Some people say that Whitney's musket parts were not truly interchangeable and that credit for the American System should go to John Hall, the New England gunsmith who built flintlock pistols for the government. Hall built many of the machine tools needed for precision manufacturing. He achieved a higher level of interchangeability and precision than did Whitney.

Oliver Evans's many inventions in the flour milling process led to an automated mill that could be run by a single miller. Samuel Colt and Elijah King Root were very successful innovators in the development of parts for the assembly-line production of firearms. Eli Terry adapted mass production methods to clock-making in the early 1800s. George Eastman made innovations in assembly-line techniques in the manufacture and developing of photographic film later in the century.

Mass production begins at Ford

Credit for the development of large-scale, assembly-line, mass production techniques is usually given to Henry Ford and his innovative

Mass production

Model T car production methods, which began in 1908. Cars were a relatively new invention and were still too expensive for the average person. Many were too heavy or low powered to be practical. Ford set out to produce a light, strong car for a reasonable price.

The methods of Henry Ford. Groups of workers at Ford initially moved down a line of parts and subassemblies, each worker carrying out a specific task. But some workers and groups were faster or slower than others, and they often got in each other's way. So Ford and his technicians decided to move the *work* instead of the *workers.*

Beginning in 1913, Ford's workers stood in one place while parts came by on conveyor belts. The Model T car moved past the workers on another conveyor belt. Car bodies were built on one line and the chassis (floor) and drive train (engine and wheels) were built on another. When both were essentially complete, the body was lowered onto the chassis for final assembly.

It has been said that Ford took the inspiration for his assembly line from the meat-processing and canning factories that moved carcasses along lines of overhead rails as early as the 1840s. Although he was not the first to use the assembly-line technique, Ford can certainly be viewed as the most successful of the early innovators due to one simple fact: Ford envisioned and fostered mass consumption as a natural consequence of mass production. His techniques lessened the time needed to build a Model T from about 12 hours to 1 hour. The price was reduced as well: from about $850 for the first Model T in 1908 to only $290 in 1927.

The mass production of chocolate-covered doughnuts. *(Reproduced by permission of The Stock Market.)*

Technique puts an end to craftsmanship

Assembly-line techniques required changing the skills necessary to build a product. Previously, each worker was responsible for the complete manufacture and assembly of all the parts needed to build any single product. This work was done by hand and relied on the individual worker's skills.

Mass production and parts interchangeability demanded that all parts be identical. Machines rather than individuality came to dictate the production process. Each part was duplicated by a machine process. The craft tradition, so important in human endeavor for centuries, was abandoned. Assembly of these machine-made parts was now divided into a series of small repetitive steps that required much less skill than traditional craftsmanship.

Modern mass production techniques changed the relationship of people to their work. Mass production has replaced craftsmanship, and the repetitive assembly line is now the world's standard for all manufacturing processes.

[*See also* **Industrial Revolution**]

Mass spectrometry

Mass spectrometry is a method for finding out the mass of particles contained in a sample and, thereby, for identifying what those particles are. A typical application of mass spectrometry is the identification of small amounts of materials found at a crime scene. Forensic (crime) scientists can use this method to identify amounts of a material too small to be identified by other means.

The basic principle on which mass spectrometry operates is that a stream of charged particles is deflected by a magnetic field. The amount of the deflection depends on the mass and the charge on the particles in the stream.

Structure of the mass spectrometer

A mass spectrometer (or mass spectrograph) consists of three essential parts: the ionization chamber, the deflection chamber, and the detector. The ionization chamber is a region in which atoms of the unknown material are excited so as to make them lose electrons. Sometimes the energy needed for exciting the atoms is obtained simply by heating the sample. When atoms are excited, they lose electrons and become positively charged particles known as ions.

Ions produced in the ionization chamber leave that chamber and pass into the deflection chamber. Their movement is controlled by an electric field whose positive charge repels the ions from the ionization chamber and whose negative charge attracts them to the deflection chamber.

The deflection chamber is surrounded by a strong magnetic field. As the stream of positive ions passes through the deflection chamber, they are deflected by the magnetic field. Instead of traveling in a straight path through the chamber, they follow a curved path. The degree to which their

Mass spectrometry

path curves is determined by the mass and charge on the positive ions. Heavier ions are not deflected very much from a straight line, while lighter ions are deflected to a greater extent.

When the positive ions leave the deflection chamber, they collide with a photographic plate or some similar material in the detector. The detector shows the extent to which particles in the unknown sample were deflected from a straight line and, therefore, the mass and charge of those particles. Since every element and every atom has a distinctive mass and charge, an observer can tell what atoms were present in the sample just by reading the record produced in the detector.

Credit for the invention of the mass spectrometer is usually given to British chemist Francis William Aston (1877–1945). Aston made a rather remarkable discovery during his first research with the mass spectrograph. When he passed a sample of pure neon gas through the instrument, he found that two separate spots showed up in the detector. The two distinct spots meant that the neon gas contained atoms of two different masses.

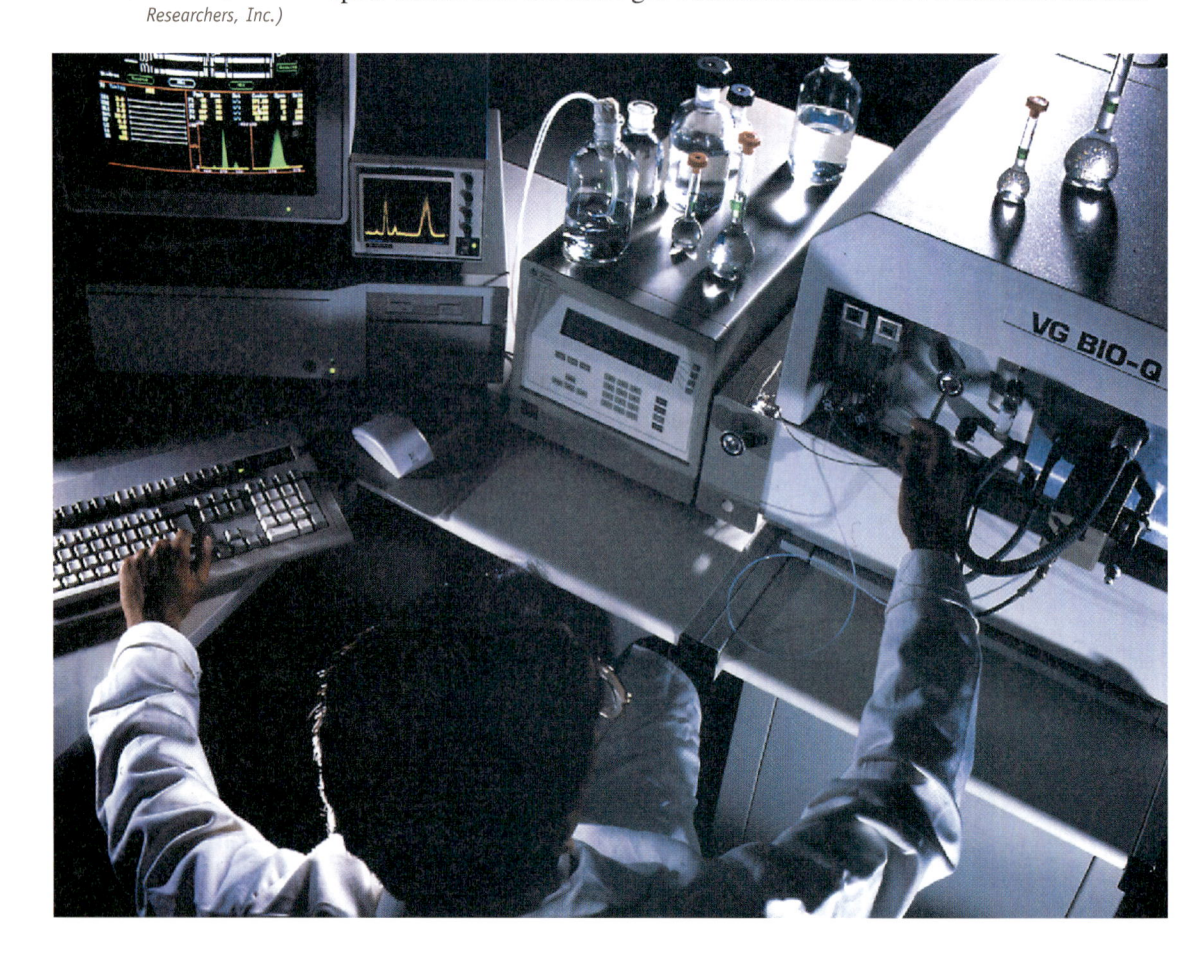

A scientist injecting a sample into a mass spectrometer. Inside, the sample will be bombarded by electrons to identify its chemical components. *(Reproduced by permission of Photo Researchers, Inc.)*

Aston interpreted this discovery to mean that two different kinds of neon atoms exist. Both atoms must have the same number of protons, since all forms of neon *always* contain the same number of protons. But the two kinds of neon atoms must have a different number of neutrons and, therefore, different atomic masses. Aston's work was the first experimental proof for the existence of isotopes, forms of the same atom that have the same number of protons but different numbers of neutrons.

[*See also* **Cathode-ray tube; Isotope**]

Mathematics

Mathematics is the science that deals with the measurement, properties, and relationships of quantities, as expressed in either numbers or symbols. For example, a farmer might decide to fence in a field and plant oats there. He would have to use mathematics to measure the size of the field, to calculate the amount of fencing needed for the field, to determine how much seed he would have to buy, and to compute the cost of that seed. Mathematics is an essential part of every aspect of life—from determining the correct tip to leave for a waiter to calculating the speed of a space probe as it leaves Earth's atmosphere.

Mathematics undoubtedly began as an entirely practical activity—measuring fields, determining the volume of liquids, counting out coins, and the like. During the golden era of Greek science, between about the sixth and third centuries B.C., however, mathematicians introduced a new concept to their study of numbers. They began to realize that numbers could be considered as abstract concepts. The number 2, for example, did not necessarily have to mean 2 cows, 2 coins, 2 women, or 2 ships. It could also represent the idea of "two-ness." Modern mathematics, then, deals both with problems involving specific, concrete, and practical number concepts (25,000 trucks, for example) and with properties of numbers themselves, separate from any practical meaning they may have (the square root of 2 is 1.4142135, for example).

Fields of mathematics

Mathematics can be subdivided into a number of special categories, each of which can be further subdivided. Probably the oldest branch of mathematics is arithmetic, the study of numbers themselves. Some of the most fascinating questions in modern mathematics involve number theory.

Mathematics

For example, how many prime numbers are there? (A prime number is a number that can be divided only by 1 and itself.) That question has fascinated mathematicians for hundreds of years. It doesn't have any particular practical significance, but it's an intriguing brainteaser in number theory.

Geometry, a second branch of mathematics, deals with shapes and spatial relationships. It also was established very early in human history because of its obvious connection with practical problems. Anyone who wants to know the distance around a circle, square, or triangle, or the space contained within a cube or a sphere has to use the techniques of geometry.

Algebra was established as mathematicians recognized the fact that real numbers (such as 4, 5.35, and $9\frac{1}{3}$) can be represented by letters. It became a way of generalizing specific numerical problems to more general situations.

Analytic geometry was founded in the early 1600s as mathematicians learned to combine algebra and geometry. Analytic geometry uses algebraic equations to represent geometric figures and is, therefore, a way of using one field of mathematics to analyze problems in a second field of mathematics.

Over time, the methods used in analytic geometry were generalized to other fields of mathematics. That general approach is now referred to as analysis, a large and growing subdivision of mathematics. One of the most powerful forms of analysis—calculus—was created almost simultaneously in the early 1700s by English physicist and mathematician Isaac Newton (1642–1727) and German mathematician Gottfried Wilhelm Leibniz (1646–1716). Calculus is a method for analyzing changing systems, such as the changes that take place as a planet, star, or space probe moves across the sky.

Statistics is a field of mathematics that grew in significance throughout the twentieth century. During that time, scientists gradually came to realize that most of the physical phenomena they study can be expressed not in terms of certainty ("A always causes B"), but in terms of probability ("A is likely to cause B with a probability of XX%"). In order to analyze these phenomena, then, they needed to use statistics, the field of mathematics that analyzes the probability with which certain events will occur.

Each field of mathematics can be further subdivided into more specific specialties. For example, topology is the study of figures that are twisted into all kinds of bizarre shapes. It examines the properties of those figures that are retained after they have been deformed.

[*See also* **Arithmetic; Calculus; Geometry; Number theory; Trigonometry**]

Matter, states of

Matter is anything that has mass and takes up space. The term refers to all real objects in the natural world, such as marbles, rocks, ice crystals, oxygen gas, water, hair, and cabbage. The term *states of matter* refers to the four physical forms in which matter can occur: solid, liquid, gaseous, and plasma.

The kinetic theory of matter

Our understanding of the nature of matter is based on certain assumptions about the particles of which matter is composed and the properties of those particles. This understanding is summarized in the kinetic theory of matter.

According to the kinetic theory of matter, all matter is composed of tiny particles. These particles can be atoms, molecules, ions, or some combination of these basic particles. Therefore, if it were possible to look

The solid, liquid, and gas states of bromine contained in a laboratory vessel. (Reproduced by permission of Photo Researchers, Inc.)

Matter, states of

Liquid Crystals

Solid, liquid, and gas: these are the three most common forms of matter. But some materials do not fit neatly into one of these three categories. Liquid crystals are one such form of matter.

Liquid crystals are materials that have properties of both solids and liquids. They exist at a relatively narrow range of temperatures. At temperatures below this range, liquid crystals act like solids. At temperatures above the range, they act like liquids.

The behavior of liquid crystals is due to the shape of the molecules of which they are made. You can think of those molecules as looking like cigars or pencils: they tend to be long and thin. They can be arranged within a material in one of three forms. In nematic crystals, the molecules are all parallel to each other but are free to move back and forth with relation to each other. The molecules in smectic crystals are also parallel to each other but they do not move back and forth; they are, however, further arranged in layers that do pass over each other. And the molecules in cholesteric crystals occur in highly structured layers that are set at slightly different angles than the ones above and below them.

at the tiniest units of which a piece of aluminum metal is composed, one would be able to observe aluminum atoms. Similarly, the smallest unit of a sugar crystal is thought to be a molecule of sugar.

The fundamental particles of which matter is composed are always in motion. Those particles may rotate on their own axes, vibrate back and forth around a certain definite point, travel through space like bullets, or display all three kinds of motion. The various states of matter differ from each other on the basis of their motion. In general, the particles of which solids are made move very slowly, liquid particles move more rapidly, and gaseous particles move much more rapidly than either solid or liquid particles. The particles of which a plasma are made have special properties that will be described later.

The motion of the particles of matter is a function of the energy they contain. Suppose that you add heat, a form of energy, to a solid. That heat is used to increase the speed with which the solid particles are moving. If enough heat is added, the particles eventually move rapidly enough that the substance turns into a liquid: it melts.

Matter, states of

The interesting property about liquid crystals is the way they transmit light. Light can pass through a liquid crystal more easily in one direction than in another. If you look at one of the crystals from one direction, you might see all the light passing through it. But from another direction, no light would be visible. The crystal would be dark.

The arrangement of molecules in a liquid crystal can be changed by adding energy to the crystal. If you warm the crystal, for example, molecules may change their position with relation to each other. This fact is utilized in new kinds of medical thermometers that change color with temperature. As body heat changes, the molecules in the liquid crystal change, the light they transmit changes, and different colors appear.

Liquid crystals are also used for the display in electronic calculators. When you press a button on the calculator, you send an electric current through the display. The electric current causes molecules in the liquid crystal to change their positions. More or less light passes through the crystal, and numbers either light up or go dark.

How states of matter differ from each other

One can distinguish among solids, liquids, and gases on two levels: the macroscopic and submicroscopic. The term macroscopic refers to properties that can be observed by the five human senses, aided or unaided. The term submicroscopic refers to properties that are too small to be seen even with the very best of microscopes.

On the macroscopic level, solids, liquids, and gases can be distinguished from each other on the basis of shape and volume. That is, solids have both constant shape and constant volume. A cube of sugar always looks exactly the same as long as it is not melted, dissolved, or changed in some other way.

Liquids have constant volume but indefinite shape. Take 100 milliliters of water in a wide pan and pour it into a tall, thin container. The total volume of the water remains the same, 100 milliliters, but the shape it takes changes.

Finally, gases have neither constant volume nor constant shape. They take the size and shape of whatever container they are placed into. Suppose you have a small container of compressed oxygen in a one-liter tank. The volume of the gas is one liter, and its shape is cylindrical (the shape of the tank). If you open the valve of the tank inside a closed room, the gas escapes to fill the room. Its volume is now much greater than 1 liter, and its shape is the shape of the room.

These macroscopic differences among solids, liquids, and gases reflect properties of the particles of which they are made. In solids, those particles are moving very slowly and tend to exert strong forces of attraction on each other. Since they have little tendency to pull away from each other, they remain in the same shape and volume.

The particles of a liquid are moving more rapidly, but they still exert a significant force on each other. These particles have the ability to flow past each other but not to escape from the attraction they feel for each other.

The particles of a gas are moving very rapidly and feel very little attraction for each other. They fly off in every direction, preventing the gas from taking on either definite shape or volume.

Plasma

Plasma is considered to be the fourth state of matter. Plasmas have been well studied in only the last few decades. They rarely exist on Earth, although they occur commonly in stars and other parts of the universe.

A plasma is a gaslike mixture with a very high temperature. The temperature of the plasma is so high that the atoms of which it is made are completely ionized. That means that the electrons that normally occur in an atom have been stripped away by the high temperature and exist independently of the atoms from which they came. A plasma is, therefore, a very hot mixture of electrons and positive ions, the atoms that are left after their electrons have been removed.

[*See also* **Atom; Crystal; Element, chemical; Gases, properties of; Ionization; Mass; Molecule**]

Mendelian laws of inheritance

Mendelian laws of inheritance are statements about the way certain characteristics are transmitted from one generation to another in an organism.

Mendelian laws
of inheritance

> **Words to Know**
>
> **Allele:** One of two or more forms a gene may take.
>
> **Dominant:** An allele whose expression overpowers the effect of a second form of the same gene.
>
> **Gamete:** A reproductive cell.
>
> **Heterozygous:** A condition in which two alleles for a given gene are different from each other.
>
> **Homozygous:** A condition in which two alleles for a given gene are the same.
>
> **Recessive:** An allele whose effects are concealed in offspring by the dominant allele in the pair.

The laws were derived by the Austrian monk Gregor Mendel (1822–1884) based on experiments he conducted in the period from about 1857 to 1865. For his experiments, Mendel used ordinary pea plants. Among the traits that Mendel studied were the color of a plant's flowers, their location on the plant, the shape and color of pea pods, the shape and color of seeds, and the length of plant stems.

Mendel's approach was to transfer pollen (which contains male sex cells) from the stamen (the male reproductive organ) of one pea plant to the pistil (female reproductive organ) of a second pea plant. As a simple example of this kind of experiment, suppose that one takes pollen from a pea plant with red flowers and uses it to fertilize a pea plant with white flowers. What Mendel wanted to know is what color the flowers would be in the offspring of these two plants. In a second series of experiments, Mendel studied the changes that occurred in the second generation. That is, suppose two offspring of the red/white mating ("cross") are themselves mated. What color will the flowers be in this second generation of plants? As a result of these experiments, Mendel was able to state three generalizations about the way characteristics are transmitted from one generation to the next in pea plants.

Terminology

Before reviewing these three laws, it will be helpful to define some of the terms used in talking about Mendel's laws of inheritance. Most of

Mendelian laws of inheritance

these terms were invented not by Mendel, but by biologists some years after his research was originally published.

Genes are the units in which characteristics are passed from one generation to the next. For example, a plant with red flowers must carry a gene for that characteristic.

A gene for any given characteristic may occur in one of two forms, called the alleles (pronounced uh-LEELZ) of that gene. For example, the gene for color in pea plants can occur in the form (allele) for a white flower or in the form (allele) for a red color.

The first step that takes place in reproduction is for the sex cells in plants to divide into two halves, called gametes. The next step is for the gametes from the male plant to combine with the gametes of the female plant to produce a fertilized egg. That fertilized egg is called a zygote. A zygote contains genetic information from both parents.

For example, a zygote might contain one allele for white flowers and one allele for red flowers. The plant that develops from that zygote would said to be heterozygous for that trait since its gene for flower color has two different alleles. If the zygote contains a gene with two identical alleles, it is said to be homozygous.

Mendel's Law of Segregation. *(Reproduced by permission of The Gale Group.)*

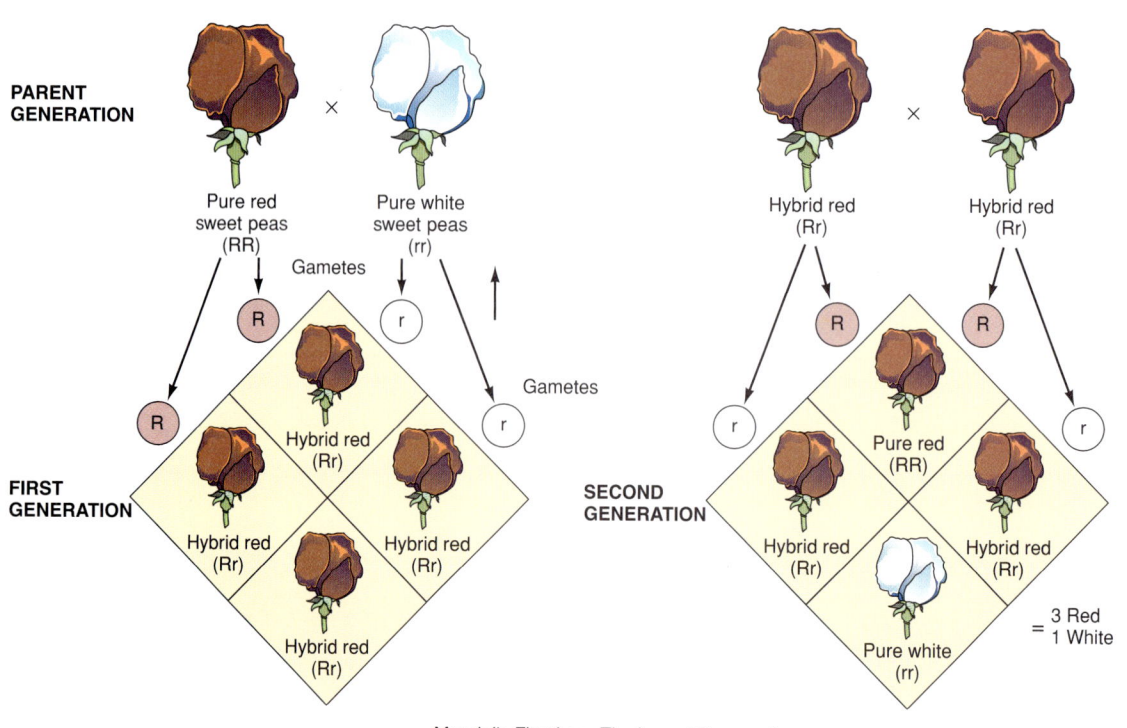

Mendel's First Law: The Law of Segregation

Mendel's laws

Mendelian laws of inheritance

Mendel's law of segregation describes what happens to the alleles that make up a gene during formation of gametes. For example, suppose that a pea plant contains a gene for flower color in which both alleles code for red. One way to represent that condition is to write RR, which indicates that both alleles (R and R) code for the color red. Another gene might have a different combination of alleles, as in Rr. In this case, the symbol R stands for red color and the r for "not red" or, in this case, white. Mendel's law of segregation says that the alleles that make up a gene separate from each other, or segregate, during the formation of gametes. That fact can be represented by simple equations, such as:

$$RR \rightarrow R + R \text{ or } Rr \rightarrow R + r$$

Mendel's second law is called the law of independent assortment. That law refers to the fact that any plant contains many different kinds of genes. One gene determines flower color, a second gene determines length of stem, a third gene determines shape of pea pods, and so on. Mendel discovered that the way in which alleles from different genes separate and then recombine is unconnected to other genes. That is, suppose that a plant contains genes for color (RR) and for shape of pod (TT). Then Mendel's second law says that the two genes will segregate independently, as:

$$RR \rightarrow R + R \text{ and } TT \rightarrow T + T$$

Mendel's third law deals with the matter of dominance. Suppose that a gene contains an allele for red color (R) and an allele for white color (r). What will be the color of the flowers produced on this plant? Mendel's answer was that in every pair of alleles, one is more likely to be expressed than the other. In other words, one allele is dominant and the other allele is recessive. In the example of an Rr gene, the flowers produced will be red because the allele R is dominant over the allele r.

Predicting traits

The application of Mendel's three laws makes it possible to predict the characteristics of offspring produced by parents of known genetic composition. The picture on page 1248, for example, shows the cross between a sweet pea plant with red flowers (RR) and one with white flowers (rr). Notice that the genes from the two parents will segregate to produce the corresponding alleles:

$$RR \rightarrow R + R \text{ and } rr \rightarrow r + r$$

There are, then, four ways in which those alleles can recombine, as shown in the same picture. However, all four combinations produce the same result: R + r → Rr. In every case, the gene formed will consist of an allele for red (R) and an allele for "not red" (r).

The drawing at the right in the picture on page 1248 shows what happens when two plants from the first generation are crossed with each other. Again, the alleles of each plant separate from each other:

$$Rr \rightarrow R + r$$

Again, the alleles can recombine in four ways. In this case, however, the results are different from those in the first generation. The possible results of these combinations are two Rr combinations, one RR combination, and one rr combination. Since R is dominant over r, three of the four combinations will produce plants with red flowers and one (the rr option) will product plants with non-red (white) flowers.

Biologists have discovered that Mendel's laws are simplifications of processes that are sometimes much more complex than the examples given here. However, those laws still form an important foundation for the science of genetics.

[See also **Chromosome; Genetics**]

Mercury (planet)

Mercury, the closest object to the Sun, is a small, bleak planet. Because of the Sun's intense glare, it is difficult to observe Mercury from Earth. Mercury is visible just above the horizon for only about one hour before sunrise and one hour after sunset.

Mercury is named for the Roman messenger god with winged sandals. The planet was so named because it orbits the Sun quickly, in just 88 days. In contrast to its short year, Mercury has an extremely long day. It takes the planet the equivalent of 59 Earth days to complete one rotation.

Mercury is the second smallest planet in the solar system (only Pluto is smaller). Mercury's diameter is about 3,000 miles (4,800 kilometers), yet it has just 5.5 percent of Earth's mass. (Earth's diameter is about 7,900 miles [12,720 kilometers].) On average, Mercury is 36 million miles (58 million kilometers) from the Sun. The Sun's intense gravitational field tilts Mercury's orbit and stretches it into a long ellipse (oval).

The *Mariner* exploration

Little else was known about Mercury until the U.S. space probe *Mariner 10* photographed the planet in 1975. *Mariner* first approached the planet Venus in February 1974, then used that planet's gravitational field to send it around like a slingshot in the direction of Mercury. The second leg of the journey to Mercury took seven weeks.

On its first flight past Mercury, *Mariner 10* came within 470 miles (756 kilometers) of the planet and photographed about 40 percent of its surface. The probe then went into orbit around the Sun and flew past Mercury twice more in the next year before running out of fuel.

Mariner 10 collected much valuable information about Mercury. It found that the planet's surface is covered with deep craters, separated by plains and huge banks of cliffs. Mercury's most notable feature is an ancient crater called the Caloris Basin, about the size of the state of Texas.

Astronomers believe that Mercury, like the Moon, was originally made of liquid rock that solidified as the planet cooled. Some meteorites hit the planet during its cooling stage and formed craters. Other meteorites,

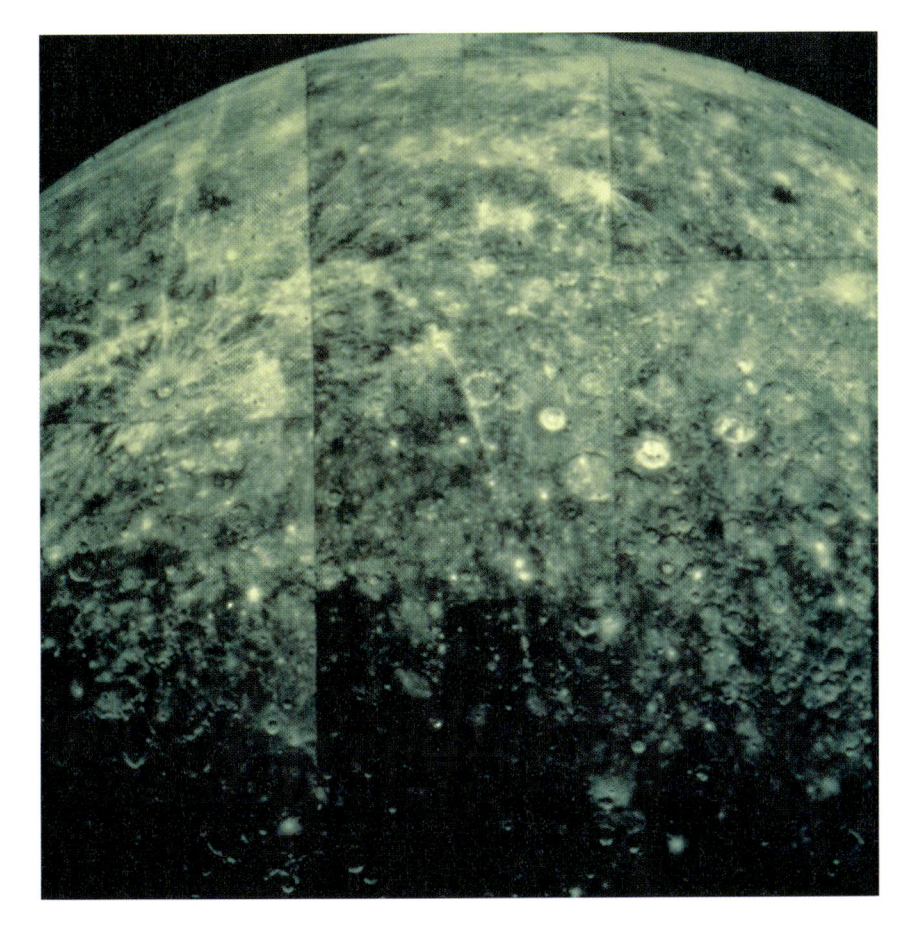

Mercury (planet)

The heavily cratered face of Mercury as seen by *Mariner 10*. Mercury shows evidence of being bombarded by meteorites throughout its history. Its largest crater is the size of the state of Texas. *(Reproduced by permission of National Aeronautics and Space Administration.)*

Mercury (planet)

On Mercury, the plains between craters, such as these located near the planet's south pole, are crossed by numerous ridges and cliffs that are similar in scale to those on Earth. (Reproduced by permission of National Aeronautics and Space Administration.)

however, broke through the cooling crust, causing lava to flow up to the surface and cover older craters, forming the plains.

Mercury's very thin atmosphere is made of sodium, potassium, helium, and hydrogen. Temperatures on Mercury reach 800°F (427°C) during its long day and −278°F (−173°C) during its long night. This temperature variation, the largest experienced by any planet in the solar system, is due to the fact that Mercury has essentially no insulating atmosphere to transport the Sun's heat from the day side to the night side.

Mariner 10 also gathered information about Mercury's core, which is nearly solid metal and is composed primarily of iron and nickel. This core, the densest of any in the solar system, accounts for about four-fifths of Mercury's diameter. It may also be responsible for creating the magnetic field that protects Mercury from the Sun's harsh particle wind.

Discovery of water on Mercury

Perhaps one of the most surprising discoveries in recent times was that of ice at Mercury's poles. The finding was made in 1991 when scientists bounced powerful radar signals off the planet's surface. Scientists had previously believed that any form of water on Mercury would rapidly evaporate given the planet's high daytime temperatures.

The polar regions of Mercury are never fully illuminated by the Sun, and it appears that ice managed to collect in the permanently shadowed regions of many polar crater rims. It is not clear where the ice came from, but scientists believe comet crashes may be one source.

Future exploration

In 2004, the National Aeronautics and Space Administration (NASA) plans to launch the $286 million MESSENGER (Mercury Surface, Space Environment, Geochemistry, and Ranging) spacecraft. It will reach Mercury five years later, enter orbit, then examine the planet's atmosphere and entire surface for one Earth year with a suite of detectors including cameras, spectrometers, and a magnetometer. MESSENGER will also explore Mercury's atmosphere and determine the size of the planet's core and how much of it is solid. Finally, the spacecraft will try to confirm whether water ice exists in polar craters on Mercury.

The European Space Agency also has ambitious plans to explore Mercury. At some future date, it proposes to send a trio of spacecraft called BepiColombo that, like MESSENGER, will study the planet's

Metabolic disorders

atmosphere and search for water ice in polar craters. BepiColombo will include two satellites and a vehicle that will land on the surface, deploying a tiny, tethered rover to gather information.

Metabolic disorders

How are your enzymes working today? Enzymes are chemical compounds that increase the rate at which reactions take place in a living organism. Without enzymes, most chemical changes in an organism would proceed so slowly that the organism could not survive. As an example, all of the metabolic reactions that take place in the body are made possible by the presence of specific enzymes. As a group these chemical reactions are referred to as metabolism.

So what happens if an enzyme is missing from the body or not functioning as it should? In such cases, a metabolic disorder may develop.

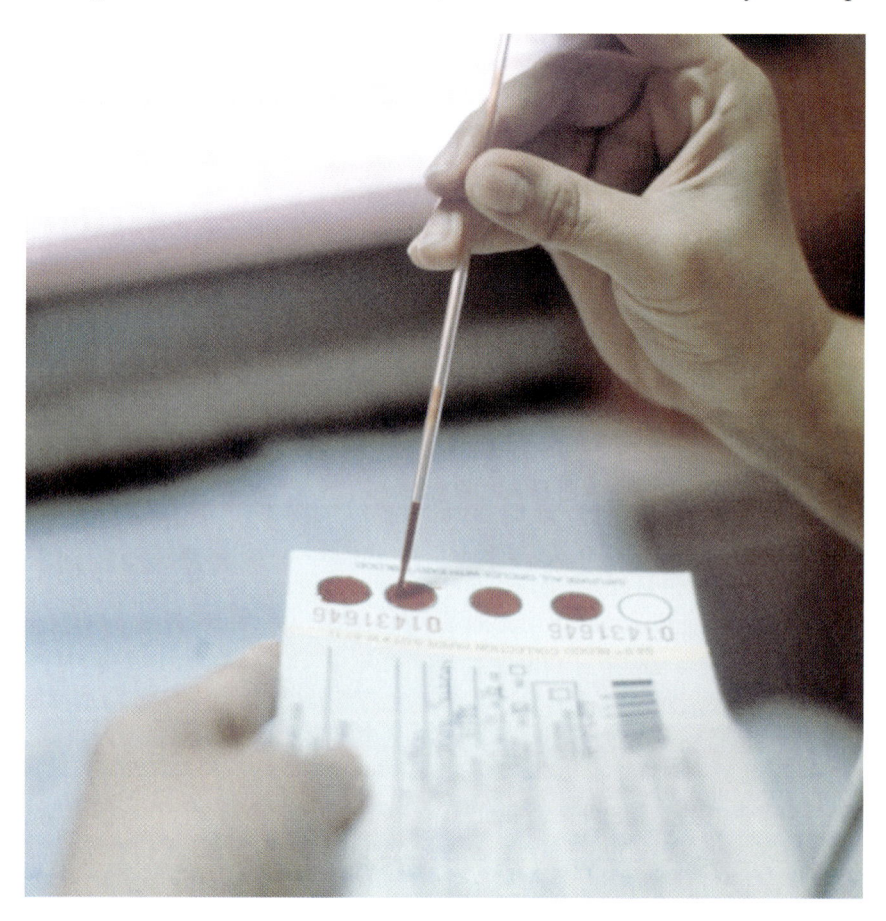

A technician performing a test for phenylketonuria (PKU). *(Reproduced by permission of Custom Medical Stock Photo, Inc.)*

A metabolic disorder is a medical condition that develops when some metabolic reaction essential for normal growth and development does not occur.

The disorder known as phenylketonuria (PKU) is an example. PKU is caused by the lack of an enzyme known as phenylalanine hydroxylase. This enzyme is responsible for converting the amino acid phenylalanine to a second amino acid, tyrosine. Tyrosine is involved in the production of the pigment melanin in the skin. Individuals with PKU are unable to make melanin and are, therefore, usually blond haired and blue eyed.

But PKU has more serious effects than light hair and eye color. When phenylalanine is *not* converted to tyrosine, it builds up in the body and is converted instead to a compound known as phenylpyruvate. Phenylpyruvate impairs normal brain development, resulting in severe mental retardation in a person with PKU. The worst symptoms of PKU can be prevented if the disorder is diagnosed early in life. In that case, a person can avoid eating foods that contain phenylalanine and developing the disorder that would follow.

Other examples of metabolic disorders include alkaptonuria, thalassemia, porphyria, Tay-Sachs disease, Hurler's syndrome, Gaucher's disease, galactosemia, Cushing's syndrome, diabetes mellitus, hyperthyroidism, and hypothyroidism. At present, no cures for metabolic disorders are available. The best approach is to diagnose such conditions as early as possible and then to arrange a person's diet to deal as effectively as possible with that disorder. Gene therapy appears to have some long-term promise for treating metabolic disorders. In this procedure, scientists attempt to provide those with metabolic disorders with the genes responsible for the enzymes they are missing, thus curing the disorder.

[*See also* **Metabolism**]

Metabolism

Metabolism refers to all of the chemical reactions that take place within an organism by which complex molecules are broken down to produce energy and by which energy is used to build up complex molecules. An example of a metabolic reaction is the one that takes place when a person eats a spoonful of sugar. Once inside the body, sugar molecules are broken down into simpler molecules with the release of energy. That energy is then used by the body for a variety of purposes, such as keeping the body warm and building up new molecules within the body.

Metabolism

> ## Words to Know
>
> **Anabolism:** The process by which energy is used to build up complex molecules.
>
> **ATP (adenosine triphosphate):** A molecule used by cells to store energy.
>
> **Carbohydrate:** A compound consisting of carbon, hydrogen, and oxygen found in plants and used as a food by humans and other animals.
>
> **Catabolism:** The process by which large molecules are broken down into smaller ones with the release of energy.
>
> **Chemical bond:** A force of attraction between two atoms.
>
> **Enzyme:** Chemical compounds that act as catalysts, increasing the rate at which reactions take place in a living organism.
>
> **Metabolic pool:** The total amount of simple molecules formed by the breakdown of nutrients.
>
> **Nutrient:** A substance that helps an organism stay alive, remain healthy, and grow.
>
> **Protein:** Large molecules that are essential to the structure and functioning of all living cells.

All metabolic reactions can be broken down into one of two general categories: catabolic and anabolic reactions. Catabolism is the process by which large molecules are broken down into smaller ones with the release of energy. Anabolism is the process by which energy is used to build up complex molecules needed by the body to maintain itself and develop.

The process of digestion

One way to understand the process of metabolism is to follow the path of a typical nutrient as it passes through the body. A nutrient is any substance that helps an organism stay alive, remain healthy, and grow. Three large categories of nutrients are carbohydrates, proteins, and fats.

Assume, for example, that a person has just eaten a piece of bread. An important nutrient in that bread is starch, a complex carbohydrate. As soon as the bread enters a person's mouth, digestion begins to occur. Enzymes in the mouth start to break down molecules of starch and convert them into smaller molecules of simpler substances: sugars. This process can be observed easily, since anyone who holds a piece of bread in his or her mouth for a period of time begins to recognize a sweet taste, the taste of the sugar formed from the breakdown of starch.

Digestion is a necessary first step for all foods. The molecules of which foods are made are too large to pass through the lining of the digestive system. Digestion results in the formation of smaller molecules that *are* able to pass through that lining and enter the person's bloodstream. Sugar molecules formed by the digestion of starch enter the bloodstream. Then they are carried to individual cells throughout a person's body.

The smaller molecules into which nutrients are broken down make up the metabolic pool. The metabolic pool consists of the simpler substances formed by the breakdown of nutrients. It includes simple sugars (formed by the breakdown of complex carbohydrates), glycerol and fatty acids (formed by the breakdown of lipids), and amino acids (formed by the breakdown of proteins). Cells use substances in the metabolic pool as building materials, just as a carpenter uses wood, nails, glue, staples, and other materials for the construction of a house. The difference is, of course, that cells construct body parts, not houses, from the materials with which they have to work.

Computer graphic of amino acid. *(Reproduced by permission of Photo Researchers, Inc.)*

Cellular metabolism

Substances that make up the metabolic pool are transported to individual cells by the bloodstream. They pass through cell membranes and enter the cell interior. Once inside a cell, a compound undergoes further metabolism, usually in a series of chemical reactions. For example, a sugar molecule is broken down inside a cell into carbon dioxide and water, with the release of energy. But that process does not occur in a single step. Instead, it takes about two dozen separate chemical reactions to convert the sugar molecule to its final products. Each chemical reaction involves a relatively modest change in the sugar molecule, the removal of a single oxygen atom or a single hydrogen atom, for example.

The purpose of these reactions is to release energy stored in the sugar molecule. To explain that process, one must know that a sugar molecule consists of carbon, hydrogen, and oxygen atoms held together by means of chemical bonds. A chemical bond is a force of attraction between two atoms. That force of attraction is a form of energy. A sugar molecule with two dozen chemical bonds can be thought of as containing two dozen tiny units of energy. Each time a chemical bond is broken, one unit of energy is set free.

Cells have evolved remarkable methods for capturing and storing the energy released in catabolic reactions. Those methods make use of very special chemical compounds, known as energy carriers. An example of such compounds is adenosine triphosphate, generally known as ATP. ATP is formed when a simpler compound, adenosine diphosphate (ADP), combines with a phosphate group. The following equation represents that change:

$$ADP + P \rightarrow ATP$$

ADP will combine with a phosphate group, as shown here, only if energy is added to it. In cells, that energy comes from the catabolism of compounds in the metabolic pool, such as sugars, glycerol, and fatty acids. In other words:

$$\text{catabolism: sugar} \rightarrow \text{carbon dioxide} + \text{water} + \text{energy};$$
$$\text{energy from catabolism} + ADP + P \rightarrow ATP$$

The ATP molecule formed in this way, then, has taken up the energy previously stored in the sugar molecule. Whenever a cell needs energy for some process, it can obtain it from an ATP molecule.

The reverse of the process shown above also takes place inside cells. That is, energy from an ATP molecule can be used to put simpler molecules together to make more complex molecules. For example, suppose that a cell needs to repair a break in its cell wall. To do so, it will need to produce new protein molecules. Those protein molecules can be made from amino acids in the metabolic pool. A protein molecule consists of hundreds or thousands of amino acid molecules joined to each other:

Amino acid 1 + amino acid 2 + amino acid 3 + (and so on) → a protein

The energy needed to *form* all the new chemical bonds needed to hold the amino acid units together comes from ATP molecules. In other words:

energy from ATP + many amino acids → protein molecule

The reactions by which a compound is metabolized differ for various nutrients. Also, energy carriers other than ATP may be involved. For example, the compound known as nicotinamide adenine dinucleotide phosphate (NADPH) is also involved in the catabolism and anabolism of various substances. The general outline shown above, however, applies to all metabolic reactions.

Metamorphosis

Metamorphosis is a series of changes through which an organism goes in developing from an early immature stage to an adult. Most people are familiar with the process, for example, by which a butterfly or moth emerges from a chrysalis (cocoon) in its adult form or a frog or toad passes through its tadpole stage.

Metamorphosis is perhaps best known among insects and amphibians (organisms such as frogs, toads, and salamanders that can live either on land or in the water). However, the process of metamorphosis has been observed in at least 17 phyla (a primary division of the animal kingdom), including Porifera (sponges), Cnidaria (jellyfish and others), Platyhelminthes (flat worms), Mollusca (mollusks), Annelida (segmented worms), Arthropoda (insects and others), Echinodermata (sea urchins and others), and Chordata (vertebrates and others).

In addition, although the term metamorphosis is generally not applied to plants, many plants do have a developmental life cycle—called the alternation of generations—which is also characterized by a dramatic change in overall body pattern.

Metamorphosis

> ### Words to Know
>
> **Alternation of generations:** A general feature of the life cycle of many plants, characterized by the occurrence of different reproductive forms that often have very different overall body patterns.
>
> **Imago:** Adult form of an insect that develops from a larva and often has wings.
>
> **Larva:** Immature form (wormlike in insects; fishlike in amphibians) of a metamorphic animal that develops from the embryo and is very different from the adult.
>
> **Molting:** Shedding of the outer layer of an animal, as occurs during growth of insect larvae.
>
> **Pupa:** A stage in the metamorphosis of an insect during which its tissues are completely reorganized to take on their adult shape.

Forms of metamorphosis

Metamorphosis in an organism is generally classified as complete or incomplete. Complete metamorphosis involves four stages: egg, larva, pupa, and adult. Consider the sequence of these stages in an insect. After a fertilized egg is laid, a wormlike larva is hatched. The larva may look like the maggot stage of a housefly or the caterpillar stage of a butterfly or moth. It is able to live on its own and secures its own food from the surrounding environment.

After a period of time, the larva builds itself some kind of protective shell such as a cocoon. The insect within the shell, now known as a pupa, is in a resting stage. It slowly undergoes a fairly dramatic change in its body structure and appearance. The energy needed for these changes comes from food eaten and stored during the larval stage.

When the process of body reorganization has been completed, the pupa breaks out of its shell and emerges in its mature adult form, also called the imago.

Incomplete metamorphosis involves only three stages, known as egg, nymph, and adult. When the fertilized egg of an insect hatches, for example, an organism appears that looks something like the adult but is smaller in size. In many cases, winged insects have not yet developed

Metamorphosis

their wings, and they are still sexually immature. In this form, the insect is known as a nymph.

Eventually, the nymph reaches a stage of maturity at which it loses its outer skin (it molts) and takes on the appearance of an adult. These stages can be seen in a grasshopper, for example, which hatches from its egg as a nymph and then passes through a series of moltings before becoming a mature adult.

[*See also* **Amphibians; Insects**]

A butterfly chrysalis (cocoon). *(Reproduced by permission of Phototake.)*

Meteor and meteorite

Meteors, also known as "shooting stars," are fragments of extraterrestrial material or, more often, small particles of dust left behind by a comet's tail. We encounter meteors every time Earth crosses the path of a comet or the debris left behind a comet. Meteors vaporize and fizzle in the atmosphere and never reach Earth's surface. At certain times of the year, large swarms of meteors, all coming from roughly the same direction, can be seen. These are called meteor showers.

Meteorites are larger chunks of rock, metal, or both that break off an asteroid or a comet and come crashing through Earth's atmosphere to strike the surface of Earth. They vary in size from a pebble to a three-ton chunk.

Early discoveries about meteors and meteorites

Until the end of the eighteenth century, people believed that meteors and meteorites were atmospheric occurrences, like rain. Other theories held that they were debris spewed into the air by exploding volcanoes, or supernatural phenomena, like signs from angry gods.

The first breakthrough in determining the true origins of meteors and meteorites came in 1714 when English astronomer Edmond Halley (1656–1742) carefully reviewed reports of their sightings. After calculating the height and speed of the objects, he concluded they must have come from space. However, he found that other scientists were hesitant to believe this notion. For nearly the next century, they continued to believe that the phenomena were Earth-based.

The conclusive evidence to confirm Halley's theory came in 1803 when a fireball, accompanied by loud explosions, rained down two to three thousand stones on northwestern France. French Academy of Science member Jean-Baptiste Biot collected some of the fallen stones as well as reports from witnesses. After measuring the area covered by the debris and analyzing the stones' composition, Biot proved they could not have originated in Earth's atmosphere.

Later observers concluded that meteors move at speeds of several miles per second. They approach Earth from space and the "flash" of a meteor is a result of its burning up upon entering Earth's atmosphere.

In November 1833, astronomers had a chance to further their understanding of meteors when a shower of thousands of shooting stars oc-

Astroblemes

Astroblemes are large, circular craters left on Earth's surface by the impact of large objects from outer space. Such objects are usually meteorites, but some may have been comet heads or asteroids. Few of these impacts are obvious today because Earth tends to erode meteorite craters over short periods of geologic time. The term astrobleme comes from two Greek roots meaning "star wound."

One of the most studied astroblemes is Barringer Crater, a meteor crater in northern Arizona that measures 0.7 miles (1.2 kilometers) across and 590 feet (180 meters) deep. It is believed to have been produced about 25,000 years ago by a nickel-iron meteorite about the size of a large house traveling at 9 miles (15 kilometers) per second. About 100 astroblemes have been identified around the world. A number are many times larger than the Barringer Crater and are hundreds of millions of years old. The largest astrobleme is South Africa's Vredefort Ring, whose diameter spans 185 miles (298 kilometers).

---

curred. Astronomers concluded that Earth was running into the objects as they were in parallel motion, like a train moving into falling rain. A look back into astronomic records revealed that a meteor shower occurred every year in November. It looked as though Earth, as it orbited the Sun, crossed the path of a cloud of meteors every November 17th. Another shower also occurred every August.

Italian scientist Giovanni Schiaparelli (1835–1910) used this information to fit the final pieces into the puzzle. He calculated the velocity and path of the August meteors, named the Perseid meteors because they appear to radiate from a point within the constellation Perseus. He found they circled the Sun in orbits similar to those of comets. He found the same to be true of the November meteors (named the Leonid meteors because they seem to originate from within the constellation Leo). Schiaparelli concluded that the paths of comets and meteor swarms were identical. Most annual meteor showers can now be traced to the orbit of a comet that intersects Earth's orbit.

The Leonid showers, occurring every year in November, are caused by the tail of comet Tempel-Tuttle, which passes through the inner solar system every 32-33 years. Such a year was 1998. On November 17 and 18 of that year, observers on Earth saw as many as 200 meteors an hour.

Meteor and meteorite

The shower was so intense that scientists and others were worried that global telecommunications might be disrupted and space telescopes damaged or destroyed. However, careful preparation by satellite and telescope engineers prevented any major disruption or damage.

What scientists now know

Through radioactive dating techniques, scientists have determined that meteorites are about 4.5 billion years old—roughly the same age as the solar system. Some are composed of iron and nickel, two elements found in Earth's core. This piece of evidence suggests that they may be fragments left over from the formation of the solar system. Further studies have shown that the composition of meteorites matches that of asteroids, leading astronomers to believe that they may originate in the asteroid belt between Mars and Jupiter.

[*See also* **Asteroid; Comet**]

Barringer Crater, an astrobleme in northern Arizona that measures 0.7 miles (1.2 kilometers) across and 590 feet (180 meters) deep. It is believed to have been created about 25,000 years ago by a meteorite about the size of a large house traveling at 9 miles (15 kilometers) per second. *(Reproduced by permission of The Corbis Corporation [Bellevue].)*

Metric system

The metric system of measurement is an internationally agreed-upon set of units for expressing the amounts of various quantities such as length, mass, time, and temperature. As of 1994, every nation in the world has adopted the metric system, with only four exceptions: the United States, Brunei, Burma, and Yemen (which use the English units of measurement).

Because of its convenience and consistency, scientists have used the metric system of units for more than 200 years. Originally, the metric system was based on only three fundamental units: the meter for length, the kilogram for mass, and the second for time. Today, there are more than 50 officially recognized units for various scientific quantities.

Measuring units in folklore and history

Nearly all early units of size were based on the always-handy human body. In the Middle Ages, the inch is reputed to have been the length of a medieval king's first thumb joint. The yard was once defined as the distance between English king Henry I's nose and the tip of his outstretched middle finger. The origin of the foot as a unit of measurement is obvious.

Eventually, ancient "rules of thumb" gave way to more carefully defined units. The metric system was adopted in France in 1799.

The metric units

The metric system defines seven basic units: one each for length, mass, time, electric current, temperature, amount of substance, and luminous intensity. (Amount of substance refers to the number of elementary particles in a sample of matter; luminous intensity has to do with the brightness of a light source.) But only four of these seven basic quantities are in everyday use by nonscientists: length, mass, time, and temperature. Their defined units are the meter for length, the kilogram for mass, the second for time, and the degree Celsius for temperature. (The other three basic units are the ampere for electric current, the mole for amount of substance, and the candela for luminous intensity.)

The meter was originally defined in terms of Earth's size; it was supposed to be one ten-millionth of the distance from the equator to the North Pole. Since Earth is subject to geological movements, this distance does not remain the same. The modern meter, therefore, is defined in terms of how far light will travel in a given amount of time when traveling at the speed of light. The speed of light in a vacuum—186,282 miles (299,727 kilometers) per hour—is considered to be a fundamental

Metric system

Metric System

MASS AND WEIGHT

Unit	Abbreviation	Mass of Grams	U.S. Equivalent (approximate)
metric ton	t	1,000,000	1.102 short tons
kilogram	kg	1,000	2.2046 pounds
hectogram	hg	100	3.527 ounces
dekagram	dag	10	0.353 ounce
gram	g	1	0.035 ounce
decigram	dg	0.1	1.543 grains
centigram	cg	0.01	0.154 grain
milligram	mg	0.001	0.015 grain
microgram	μm	0.000001	0.000015 grain

LENGTH

Unit	Abbreviation	Mass of Grams	U.S. Equivalent (approximate)
kilometer	km	1,000	0.62 mile
hectometer	hm	100	328.08 feet
dekameter	dam	10	32.81 feet
meter	m	1	39.37 inches
decimeter	dm	0.1	3.94 inches
centimeter	cm	0.01	0.39 inch
millimeter	mm	0.001	0.039 inch
micrometer	μm	0.000001	0.000039 inch

constant of nature that will never change. The standard meter is equivalent to 39.3701 inches.

The kilogram is the metric unit of mass, not weight. Mass is the fundamental measure of the amount of matter in an object. Unfortunately, no absolutely unchangeable standard of mass has yet been found on which to standardize the kilogram. The kilogram is therefore defined as the mass of a certain bar of platinum-iridium alloy that has been kept since 1889 at the International Bureau of Weights and Measures in Sèvres, France. The kilogram is equivalent to 2.2046 pounds.

The metric unit of time is the same second that has always been used, except that it is now defined in a very accurate way. It no longer depends on the wobbly rotation of our planet (1/86,400th of a day), be-

AREA

Unit	Abbreviation	Mass of Grams	U.S. Equivalent (approximate)
square kilometer	sq km or km²	1,000,000	0.3861 square miles
hectare	ha	10,000	2.47 acres
are	a	100	119.60 square yards
square centimeter	sq cm or cm²	0.0001	0.155 square inch

VOLUME

Unit	Abbreviation	Mass of Grams	U.S. Equivalent (approximate)
cubic meter	m³	1	1.307 cubic yards
cubic decimeter	dm³	0.001	61.023 cubic inches
cubic centimeter	cu cm or cm³ or cc	0.000001	0.061 cubic inch

CAPACITY

Unit	Abbreviation	Mass of Grams	U.S. Equivalent (approximate)
kiloliter	kl	1,000	1.31 cubic yards
hectoliter	hl	100	3.53 cubic feet
dekaliter	dal	10	0.35 cubic foot
liter	l	1	61.02 cubic inches
cubic decimeter	dm³	1	61.02 cubic inches
deciliter	dl	0.10	6.1 cubic inches
centiliter	cl	0.01	0.61 cubic inch
milliliter	ml	0.001	0.061 cubic inch
microliter	µl	0.000001	0.000061 cubic inch

cause Earth is slowing down. Days keep getting a little longer as Earth grows older. So the second is now defined in terms of the vibrations of a certain kind of atom known as cesium-133. One second is defined as the amount of time it takes for a cesium-133 atom to vibrate in a particular way 9,192,631,770 times. Because the vibrations of atoms depend only on the nature of the atoms themselves, cesium atoms will presumably continue to behave exactly like cesium atoms forever. The exact number of cesium vibrations was chosen to come out as close as possible to what was previously the most accurate value of the second.

The metric unit of temperature is the degree Celsius, which replaces the English system's degree Fahrenheit. It is impossible to convert between Celsius and Fahrenheit simply by multiplying or dividing by 1.8,

Microwave communication

however, because the scales start at different places. That is, their zero-degree marks have been set at different temperatures.

Bigger and smaller metric units

In the metric system, there is only one basic unit for each type of quantity. Smaller and larger units of those quantities are all based on powers of ten (unlike the English system that invents different-sized units with completely different names based on different conversion factors: 3, 12, 1760, etc.). To create those various units, the metric system simply attaches a prefix to the name of the unit. Latin prefixes are added for smaller units, and Greek prefixes are added for larger units. The basic prefixes are: kilo- (1000), hecto- (100), deka- (10), deci- (0.1), centi- (0.01), and milli- (0.001). Therefore, a kilometer is 1,000 meters. Similarly, a millimeter is one-thousandth of a meter.

Minutes are permitted to remain in the metric system even though they don't conform strictly to the rules. The minute, hour, and day, for example, are so customary that they're still defined in the metric system as 60 seconds, 60 minutes, and 24 hours—not as multiples of ten. For volume, the most common metric unit is not the cubic meter, which is generally too big to be useful in commerce, but the liter, which is one-thousandth of a cubic meter. For even smaller volumes, the milliliter, one-thousandth of a liter, is commonly used. And for large masses, the metric ton is often used instead of the kilogram. A metric ton (often spelled tonne) is 1,000 kilograms. Because a kilogram is about 2.2 pounds, a metric ton is about 2,200 pounds: 10 percent heavier than an American ton of 2,000 pounds. Another often-used, nonstandard metric unit is the hectare for land area. A hectare is 10,000 square meters and is equivalent to 0.4047 acre.

[*See also* **Units and standards**]

Microwave communication

A microwave is an electromagnetic wave with a very short wavelength, between .039 inches (1 millimeter) and 1 foot (30 centimeters). Within the electromagnetic spectrum, microwaves can be found between radio waves and shorter infrared waves. Their short wavelengths make microwaves ideal for use in radio and television broadcasting. They can transmit along a vast range of frequencies without causing signal interference or overlap.

Words to Know

Electromagnetic radiation: Radiation that transmits energy through the interaction of electricity and magnetism.

Electromagnetic spectrum: The complete array of electromagnetic radiation, including radio waves (at the longest-wavelength end), microwaves, infrared radiation, visible light, ultraviolet radiation, X rays, and gamma rays (at the shortest-wavelength end).

Microwave technology was developed during World War II (1939–45) in connection with secret military radar research. Today, microwaves are used primarily in microwave ovens and communications. A microwave communications circuit can transmit any type of information as efficiently as telephone wires.

The most popular devices for generating microwaves are magnetrons and klystrons. They produce microwaves of low power and require the use of an amplification device, such as a maser (microwave amplification by stimulated emission of radiation). Like radio waves, microwaves can be modulated for communication purposes. However, they offer 100 times more useful frequencies than radio.

Microwaves can be easily broadcast and received via aerial antennas. Unlike radio waves, microwave signals can be focused by antennas just as a searchlight concentrates light into a narrow beam. Signals are transmitted directly from a source to a receiver site. Reliable microwave signal range does not extend very far beyond the visible horizon.

It is standard practice to locate microwave receivers and transmitters atop high buildings when hilltops or mountain peaks are not available. The higher the antenna, the farther the signal can be broadcast. It takes many ground-based relay "hops" to carry a microwave signal across a continent. Since the 1960s, the United States has been spanned by a network of microwave relay stations.

A more common method of microwave transmission is the waveguide. Waveguides are hollow pipes that conduct microwaves along their inner walls. They are constructed from materials of very high electric conductivity and must be of precise design. Waveguides operate only at very high frequencies, so they are ideal microwave conductors.

Microwave communication

Satellites and microwaves

Earth satellites relaying microwave signals from the ground have increased the distance that can be covered in one hop. Microwave repeaters in a satellite in a stationary orbit 22,300 miles (35,880 kilometers) above Earth can reach one-third of Earth's surface. More than one-half of the long-distance phone calls made in the United States are routed through satellites via microwaves.

A microwave communications tower in Munich, Germany. (Reproduced by permission of Photo Researchers, Inc.)

The weather and microwave communication

Raindrops and hailstones are similar in size to the wavelength of higher-frequency microwaves. A rainstorm can block microwave communication, producing a condition called rain fade. To locate incoming storms, weather radar deliberately uses shorter-wavelength microwaves to increase interaction with rain.

Microwave communication is nearly 100 percent reliable. The reason is that microwave communication circuits have been engineered to minimize fading, and computer-controlled networks often reroute signals through a different path before a fade becomes noticeable.

[*See also* **Antenna**]

Migration

In biology, the term migration refers to the regular, periodic movement of animals between two different places. Migration usually occurs in response to seasonal changes and is motivated by breeding and/or feeding drives. Migration has been studied most intensively among birds, but it is known to take place in many other animals as well, including insects, fish, whales, and other mammals. Migration is a complex behavior that involves timing, navigation, and other survival skills.

The term migration also applies to the movement of humans from one country to another for the purpose of taking up long-term or permanent residency in the new country.

Types of migration

Four major types of migration are known. In complete migration, all members of a population travel from their breeding habitat at the end of that season, often to a wintering site hundreds or even thousands of kilometers away. The arctic tern is an example of a complete migrant. Individuals of this species travel from the Arctic to the Antarctic and back again during the course of a year, a round-trip migration of more than 30,000 kilometers!

In other species, some individuals remain at the breeding ground year-round while other members of the same species migrate away. This phenomenon is known as partial migration. American robins are considered indicators of the arrival of spring in some areas but are year-round residents in other areas.

Migration

Differential migration occurs when all the members of a population migrate, but not necessarily at the same time or for the same distance. The differences are often based on age or sex. Herring gulls, for example, migrate a shorter and shorter distance as they grow older. Male American kestrels spend more time at their breeding grounds than do females, and when they do migrate, they don't travel as far.

Irruptive migration occurs in species that do not migrate at all during some years but may do so during other years. The primary factors determining whether or not migration occurs are weather and availability of food. For example, some populations of blue jays are believed to migrate only when their winter food of acorns is scarce.

Migration Pathways

Migratory animals travel along the same general routes each year. Several common "flyways" are used by North American birds on their southward journey. The most commonly used path includes an 800 to 1,100 kilometer flight southward across the Gulf of Mexico. In order to survive this difficult journey, birds must store extra energy in the form of fat. All along the migration route, but particularly before crossing a large expanse of water, birds rest and eat, sometimes for days at a time. The

Caribou in the Arctic National Wildlife Refuge. Some caribou migrate more than 600 miles (965 kilometers) to spend the winter in forests. *(Reproduced by permission of the U.S. Fish and Wildlife Service.)*

birds start out again on their journey only when they have added a certain amount of body fat.

Although most migrants travel at night, a few birds prefer daytime migrations. The pathways used by these birds tend to be less direct and slower than those of night migrants, primarily because of differences in feeding strategies. Night migrants can spend the day in one area foraging for food and building up energy reserves for the night's nonstop flight. Daytime migrants must combine travel with foraging, and thus tend to keep to the shorelines, which are rich in insect life, capturing food during a slow but ever-southward journey.

Navigation

Perhaps the most remarkable aspect of migration is the navigational skills employed by the animals. Birds such as the albatross and lesser golden plover travel hundreds of kilometers over the featureless open ocean. Yet they arrive home without error to the same breeding grounds year after year. Salmon migrate upstream from the sea to the very same freshwater shallows in which they were hatched. Monarch butterflies began life in the United States or Canada. They then travel to the same wintering grounds in Southern California or Mexico that had been used by ancestors many generations before.

How are these incredible feats of navigation accomplished? Different animals have been shown to use a diverse range of navigational aids, involving senses often much more acute than our own. Sight, for example, may be important for some animals' navigational skills, although it may often be secondary to other senses. Salmon can smell the water of their home rivers, and follow this scent all the way from the sea. Pigeons also sense wind-borne odors and may be able to organize the memories of the sources of these smells in a kind of internal map. It has been shown that many animals have the ability to sense the magnetic forces associated with the north and south poles, and thus have their own built-in compass. This magnetic sense and the sense of smell are believed to be the most important factors involved in animal migration.

Minerals

Minerals are the natural, inorganic (nonliving) materials that compose rocks. Examples are gems and metals. Minerals have a fixed chemical makeup and a definite crystal structure (its atoms are arranged in orderly

Minerals

> ## Words to Know
>
> **Compound:** A substance consisting of two or more elements in specific proportions.
>
> **Crystal:** Naturally occurring solid composed of atoms or molecules arranged in an orderly pattern that repeats at regular intervals.
>
> **Element:** Pure substance composed of just one type of atom that cannot be broken down chemically into simpler substances.
>
> **Metallurgy:** Science and technology of extracting metals from their ores and refining them for use.
>
> **Ore:** Mineral compound that is mined for one of the elements it contains, usually a metal element.
>
> **Rock:** Naturally occurring solid mixture of minerals.
>
> **Silicate:** Mineral containing the elements silicon and oxygen, and usually other elements as well.

patterns). Therefore, a sample of a particular mineral will have essentially the same composition no matter where it is from—Earth, the Moon, or beyond. Properties such as crystal shape, color, hardness, density, and luster distinguish minerals from each other. The study of the distribution, identification, and properties of minerals is called mineralogy.

Almost 4,000 different minerals are known, with several dozen new minerals identified each year. However, only 20 or so minerals compose the bulk of Earth's crust, the part of Earth extending from the surface downward to a maximum depth of about 25 miles (40 kilometers). These minerals are often called the rock-forming minerals.

Mineralogists group minerals according to the chemical elements they contain. Elements are substances that are composed of just one type of atom. Over 100 of these are known, of which 88 occur naturally. Only ten elements account for nearly 99 percent of the weight of Earth's crust. Oxygen is the most plentiful element, accounting for almost 50 percent of that weight. The remaining elements are (in descending order) silicon, aluminum, iron, calcium, sodium, potassium, magnesium, hydrogen, and titanium.

Most minerals are compounds, meaning they contain two or more elements. Since oxygen and silicon together make up almost three-quarters of the mass of Earth's crust, the most abundant minerals are silicate minerals—compounds of silicon and oxygen. The major component of nearly every kind of rock, silicate compounds generally contain one or more metals, such as calcium, magnesium, aluminum, and iron.

Only a few minerals, known as native elements, contain atoms of just a single element. These include the so-called native metals: platinum, gold, silver, copper, and iron. Diamond and graphite are both naturally occurring forms of pure carbon, but their atoms are arranged differently. Sulfur, a yellow nonmetal, is sometimes found pure in underground deposits formed by hot springs.

Physical traits and mineral identification

A mineral's physical traits are a direct result of its chemical composition and crystal form. Therefore, if enough physical traits are recognized, any mineral can be identified. These traits include hardness, color, streak, luster, cleavage or fracture, and specific gravity.

Hardness. A mineral's hardness is defined as its ability to scratch another mineral. This is usually measured using a comparative scale devised in 1822 by German mineralogist Friedrich Mohs. The Mohs hardness scale lists 10 common minerals, assigning to each a hardness from 1 (talc) to 10 (diamond). A mineral can scratch all those minerals having a lower Mohs hardness number. For example, calcite (hardness 3) can scratch gypsum (hardness 2) and talc (hardness 1), but it cannot scratch fluorite (hardness 4).

Color and streak. Although some minerals can be identified by their color, this can be misleading since mineral color is often affected by traces of impurities. Streak, however, is a very reliable identifying feature. Streak refers to the color of the powder produced when a mineral is scraped across an unglazed porcelain tile called a streak plate. Fluorite, for example, comes in a great range of colors, yet its streak is always white.

Luster. Luster refers to a mineral's appearance when light reflects off its surface. There are various kinds of luster, all having descriptive names. Thus, metals have a metallic luster, quartz has a vitreous or glassy luster, and chalk has a dull or earthy luster.

Cleavage and fracture. Some minerals, when struck with force, will cleanly break along smooth planes that are parallel to each other. This breakage is called cleavage and is determined by the way a mineral's atoms are arranged. Muscovite cleaves in one direction only, producing

Minerals

thin flat sheets. Halite cleaves in three directions, all perpendicular to each other, forming cubes.

However, most minerals fracture rather than cleave. Fracture is breakage that does not follow a flat surface. Some fracture surfaces are rough and uneven. Those that break along smooth, curved surfaces like a shell are called conchoidal fractures. Breaks along fibers are called fibrous fractures.

Specific gravity. The specific gravity of a mineral is the ratio of its weight to that of an equal volume of water. Water has a specific gravity

A sample of gold leaf from Tuolomne County, California. *(Reproduced by permission of National Aeronautics and Space Administration.)*

of 1.0. When pure, each mineral has a predictable specific gravity. Most range between 2.2 and 3.2. (This means that most are 2.2 to 3.2 times as heavy as an equal volume of water.) Quartz has a specific gravity of 2.65, while the specific gravity of gold is 19.3.

Mineral resources

Everything that humankind consumes, uses, or produces has its origin in minerals. Minerals are the building materials of our technological

A sample of rose quartz wrapped around quartz from Sapucaia Pegmatite, Brazil. (Reproduced by permission of National Aeronautics and Space Administration.)

civilization, from microprocessors made of silicon to skyscrapers made of steel.

Gems or gemstones are minerals that are especially beautiful and rare. The beauty of a gem depends on its luster, color, and hardness. The so-called precious stones are diamond, ruby, sapphire, and emerald. Some semiprecious stones are amethyst, topaz, garnet, opal, turquoise, and jade. The weight of gems are measured in carats: one carat equals 200 milligrams (0.007 ounces).

Precious metals have also acquired great value because of their beauty, rarity, and durability. Platinum, gold, and silver are the world's precious metals. Other metals, although not considered precious, are commercially valuable. Examples include copper, lead, aluminum, zinc, iron, mercury, nickel, and chromium.

A mineral compound that is mined for a metal element it contains is called an ore. Metallurgy is the science and technology of extracting metals from their ores and refining them for use. Iron, which alone accounts for over 90 percent of all metals mined, is found in the ores magnetite and hematite. These ores contain 15 to 60 percent iron. Other ores, however, contain very little metal. One ton of copper ore may yield only about eight pounds of copper (one metric ton may yield only four kilograms). The remaining material is considered waste.

[*See also* **Crystal; Industrial minerals; Mining; Precious metals; Rocks**]

Mining

Mining is the process by which commercially valuable mineral resources are extracted (removed) from Earth's surface. These resources include ores (minerals usually containing metal elements), precious stones (such as diamonds), building stones (such as granite), and solid fuels (such as coal). Although many specific kinds of mining operations have been developed, they can all be classified into one of two major categories: surface and subsurface (or underground) mining.

History

Many metals occur in their native state or in readily accessible ores. Thus, the working of metals (metallurgy) actually dates much farther back than does the mining industry itself. Some of the earliest known mines were

Words to Know

Adit: A horizontal tunnel constructed to gain access to underground mineral deposits.

Metallurgy: Science and technology of extracting metals from their ores and refining them for use.

Ore: A mineral compound that is mined for one of the elements it contains, usually a metal element.

Overburden: Rocky material that must be removed in order to gain access to an ore or coal bed.

Prospecting: The act of exploring an area in search of mineral deposits or oil.

Shaft: A vertical tunnel constructed to gain access to underground mineral deposits.

those developed by the Greeks in the sixth century B.C. By the time the Roman Empire reached its peak, it had established mining sites throughout the European continent, in the British Isles, and in parts of North Africa. Some of the techniques used to shore up underground mines still in use today were introduced as far back as the Greek and Roman civilizations.

Exploration

Until the beginning of the twentieth century, prospecting (exploring an area in search of mineral resources) took place in locations where ores were readily available. During the California and Alaska gold rushes of the nineteenth century, prospectors typically found the ores they were seeking in outcrops visible to the naked eye or by separating gold and silver nuggets from stream beds. Over time, of course, the supply of these readily accessible ores was exhausted and different methods of mining were developed.

Surface mining

When an ore bed has been located relatively close to Earth's surface, it can be mined by surface techniques. Surface mining is generally a much preferred approach to mining because it is less expensive and safer

Mining

than subsurface mining. In fact, about 90 percent of the rock and mineral resources mined in the United States and more than 60 percent of the nation's coal is produced by surface mining techniques.

Surface mining can be subdivided into two large categories: open-pit mining and strip mining. Open-pit mining is used when an ore bed covers a very large area in both distance and depth. Mining begins when scrapers remove any non-ore material (called overburden) on top of the ore. Explosives are then used to blast apart the ore bed itself. Fragments from the blasting are hauled away in large trucks. As workers dig downward into the ore bed, they also expand the circular area in which they work. Over time, the open-pit mine develops the shape of a huge bowl with terraces or ledges running around its inside edge. The largest open-pit mine in the United States has a depth of more than 0.5 mile (0.8 kilometer) and a diameter of 2.25 miles (3.6 kilometers). Open-pit mining continues until the richest part of the ore bed has been excavated.

When an ore bed covers a wide area but is not very deep, strip mining is used. It begins the same as open-pit mining, with scrapers and other machines removing any overburden. This step involves the removal of two long parallel rows of material. As the second row is dug, the overburden removed is dumped into the first row. The ore exposed in the second row is then extracted. When that step has been completed, machines remove the overburden from a third parallel row, dumping the material extracted into the second row. This process continues until all the ore has been removed from the area. Afterward, the land typically resembles a washboard with parallel rows of hills and valleys consisting of excavated soil.

Subsurface mining

Ores and other mineral resources may often lie hundreds or thousands of feet beneath Earth's surface. Because of this, their extraction is difficult. To gain access to these resources, miners create either a horizontal tunnel (an adit) or a vertical tunnel (a shaft). To ensure the safety of workers, these tunnels must be reinforced with wooden timbers and ceilings. In addition, ventilation shafts must be provided to allow workers a sufficient supply of air, which is otherwise totally absent within the mine.

Once all safety procedures have been completed, the actual mining process begins. In many cases, the first step is to blast apart a portion of the ore deposit with explosives. The broken pieces obtained are then collected in carts or railroad cars and taken to the mine opening.

Other techniques for the mining of subsurface resources are also available. The removal of oil and natural gas by drilling into Earth's surface are well-known examples. Certain water-soluble minerals can be removed by dissolving them with hot water that is piped into the ground under pressure. The dissolved minerals are then carried to the surface.

Environmental issues

In general, subsurface mining is less environmentally hazardous than surface mining. One problem with subsurface mining is that underground mines sometimes collapse, resulting in the massive sinking of land above

Earth movers strip mining for coal in West Virginia. *(Reproduced by permission of The Stock Market.)*

them. Another problem is that waste materials produced during mining may be dissolved by underground water, producing water solutions that are poisonous to plant and animal life.

In many parts of the United States, vast areas of land have been laid bare by strip mining. Often, it takes many years for vegetation to start regrowing once more. Even then, the land never quite assumes the appearance it had before mining began. Strip mining also increases land erosion, resulting in the loss of soil and in the pollution of nearby waterways.

[*See also* **Coal; Minerals; Precious metals**]

Mole

In chemistry, a mole is a certain number of particles, usually of atoms or molecules. In theory, one could use any number of different terms for counting particles in chemistry. For example, one could talk about a dozen (12) particles or a gross (144) of particles. The problem with these terms is that they describe far fewer particles than one usually encounters in chemistry. Even the tiniest speck of sodium chloride (table salt), for example, contains trillions and trillions of particles.

The term mole, by contrast, refers to 6.022137×10^{23} particles. Written out in the long form, it's 602,213,700,000,000,000,000,000 particles. This number is very special in chemistry and is given the name Avogadro's number, in honor of Italian chemist and physicist Amadeo Avogadro (1776–1856), who first suggested the concept of a molecule.

A unit like the mole (abbreviated mol) is needed because of the way chemists work with and think about matter. When chemists work in the laboratory, they typically handle a few grams of a substance. They might mix 15 grams of sodium with 15 grams of chlorine. But when substances react with each other, they don't do so by weight. That is, one gram of sodium does *not* react exactly with one gram of chlorine.

Instead, substances react with each other atom-by-atom or molecule-by-molecule. In the above example, one atom of sodium combines with one atom of chlorine. This ratio is not the same as the weight ratio because one atom of sodium weighs only half as much as one atom of chlorine.

The mole unit, then, acts as a bridge between the level on which chemists actually work in the laboratory (by weight, in grams) and the way substances actually react with each other (by individual particles, such as atoms). One mole of any substance—no matter what substance it is—always contains the same number of particles: the Avogadro number of particles.

Think of what this means in the reaction between sodium and chlorine. If a chemist wants this reaction to occur completely, then exactly the same number of particles of each must be added to the mixture. That is, the same number of moles of each must be used. One can say: 1 mole of sodium will react completely with 1 mole of chlorine. It's easy to calculate a mole of sodium; it is the atomic weight of sodium (22.98977) expressed in grams. And it's easy to calculate a mole of chlorine; it is the molecular weight of chlorine (70.906) expressed in grams. This conversion allows the chemist to weigh out exactly the right amount of sodium and chlorine to make sure the reaction between the two elements goes to completion.

Molecular biology

Molecular biology is the study of life at the level of atoms and molecules. Suppose, for example, that one wishes to understand as much as possible about an earthworm. At one level, it is possible to describe the obvious characteristics of the worm, including its size, shape, color, weight, the foods it eats, and the way it reproduces.

Long ago, however, biologists discovered that a more basic understanding of any organism could be obtained by studying the cells of which that organism is made. They could identify the structures of which cells are made, the way cells change, the substances needed by the cell to survive, products made by the cell, and other cellular characteristics.

Molecular biology takes this analysis of life one step further. It attempts to study the molecules of which living organisms are made in much the same way that chemists study any other kind of molecule. For example, they try to find out the chemical structure of these molecules and the way this structure changes during various life processes, such as reproduction and growth. In their research, molecular biologists make use of ideas and tools from many different sciences, including chemistry, biology, and physics.

The Central Dogma

The key principle that dominates molecular biology is known as the Central Dogma. (A dogma is an established belief.) The Central Dogma is based on two facts. The first fact is that the key players in the way any cell operates are proteins. Proteins are very large, complex molecules made

Molecular biology

> ## Words to Know
>
> **Amino acid:** An organic compound from which proteins are made.
>
> **Cell:** The basic unit of a living organism; cells are structured to perform highly specialized functions.
>
> **Cytoplasm:** The semifluid substance of a cell containing organelles and enclosed by the cell membrane.
>
> **DNA (deoxyribonucleic acid):** The genetic material in the nucleus of cells that contains information for an organism's development.
>
> **Enzyme:** Any of numerous complex proteins that are produced by living cells and spark specific biochemical reactions.
>
> **Hormone:** A chemical produced in living cells that is carried by the blood to organs and tissues in distant parts of the body, where it regulates cellular activity.
>
> **Nucleotide:** A unit from which DNA molecules are made.
>
> **Protein:** A complex chemical compound that consists of many amino acids attached to each other that are essential to the structure and functioning of all living cells.
>
> **Ribosome:** Small structures in cells where proteins are produced.

of smaller units known as amino acids. A typical protein might consist, as an example, of a few thousand amino acid molecules joined to each other end-to-end. Proteins play a host of roles in cells. They are the building blocks from which cell structures are made; they act as hormones (chemical messengers) that deliver messages from one part of a cell to another or from one cell to another cell; and they act as enzymes, compounds that speed up the rate at which chemical reactions take place in cells.

The second basic fact is that proteins are constructed in cells based on master plans stored in molecules known as deoxyribonucleic acids (DNA) present in the nuclei of cells. DNA molecules consist of very long chains of units known as nucleotides joined to each other end-to-end. The sequence in which nucleotides are arranged act as a kind of code that tells a cell what proteins to make and how to make them.

The Central Dogma, then, is very simple and can be expressed as follows:

$$\text{DNA} \to \text{mRNA} \to \text{proteins}$$

What this equation says in words is that the code stored in DNA molecules in the nucleus of a cell is first written in another kind of molecule known as messenger ribonucleic acid (mRNA). Once they are constructed, mRNA molecules leave the nucleus and travel out of the nucleus into the cytoplasm of the cell. They attach themselves to ribosomes, structures inside the cytoplasm where protein production takes place. Amino acids that exist abundantly in the cytoplasm are then brought to the ribosomes by another kind of RNA, transfer RNA (tRNA), where they are used to construct new protein molecules. These molecules have their structure dictated by mRNA molecules which, in turn, have structures originally dictated by DNA molecules.

Significance of molecular biology

The development of molecular biology has provided a new and completely different way of understanding living organisms. We now know, for example, that the functions a cell performs can be described in chemical terms. Suppose that we know that a cell makes red hair. What we have learned is that the reason the cell makes red hair is that DNA molecules in its nucleus carry a coded message for red-hair-making. That coded message passes from the cell's DNA to its mRNA. The mRNA then directs the production of red-hair proteins.

The same can be said for any cell function. Perhaps a cell is responsible for producing antibodies against infection, or for making the hormone insulin, or assembling a sex hormone. All of these cell functions can be specified as a set of chemical reactions.

But once that fact has been realized, then humans have exciting new ways of dealing with living organisms. If the master architect of cell functions is a chemical molecule (DNA), then that molecule can be changed, like any other chemical molecule. If and when that happens, the functions performed by the cell are also changed. For these reasons, the development of molecular biology is regarded by many people as one of the greatest revolutions in all of scientific history.

Molecule

A molecule is a particle consisting of two or more atoms joined to each other by means of a covalent bond. (Electrons are shared in covalent bonds.) There are a number of different ways of representing molecules.

Molecule

> ## Words to Know
>
> **Atom:** The smallest particle of which an element can exist.
>
> **Chemical bond:** An electrical force of attraction that holds two atoms together.
>
> **Covalent bond:** A chemical bond formed when two atoms share a pair of electrons with each other.
>
> **Compound:** A substance consisting of two or more elements in specific proportions.
>
> **Element:** A pure substance that cannot be broken down into anything simpler by ordinary chemical means.
>
> **Molecular formula:** A shorthand method for representing the composition of a molecule using symbols for the type of atoms involved and subscripts for the number of atoms involved.
>
> **Molecule:** A particle formed when two or more atoms join together.
>
> **Structural formula:** The chemical representation of a molecule that shows how the atoms are arranged within the molecule.

Figure 1. Space-filling models of various elements. *(Reproduced by permission of The Gale Group.)*

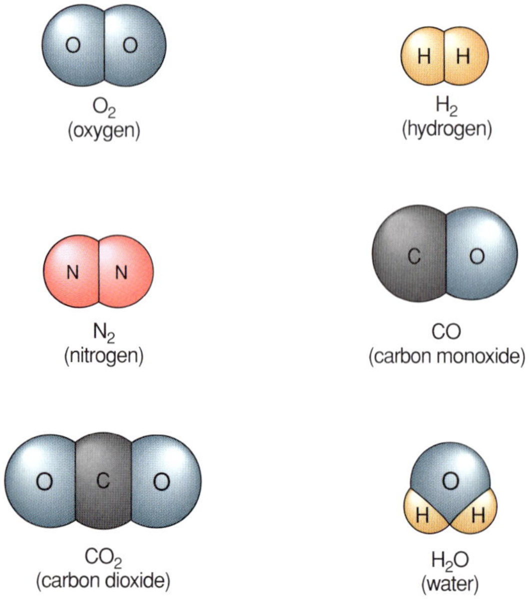

O_2 (oxygen)

H_2 (hydrogen)

N_2 (nitrogen)

CO (carbon monoxide)

CO_2 (carbon dioxide)

H_2O (water)

One method is called an electron-dot diagram, which shows the atoms included in the molecule and the electron pairs that hold the atoms together. Another method is the ball-and-stick model, in which the atoms present in the molecule are represented by billiard-ball-like spheres; the bonds that join them are represented by wooden sticks. A third method is called a space-filling model, which shows the relative size of the atoms in the molecule and the way the atoms are actually arranged in space (see Figure 1).

Formation of compounds

A compound is formed when two atoms of an element react with each other. For example, water is formed when atoms of hydrogen react with atoms of oxygen. The reaction between two atoms always involves the exchange of electrons between the two atoms. One atom tends to lose one or more electrons, and the other atom tends to gain that (or those) electrons.

In general, this exchange of electrons can occur in two ways. First, one atom can completely lose its electrons to the second atom. The first atom, with fewer electrons than usual, becomes a positively charged particle called a cation. The second, with more electrons than usual, becomes a negatively charged particle called an anion. A compound formed in this way consists of pairs of ions, some positive and some negative. The ions stay together because they carry opposite electric charges, and opposite electric charges attract each other.

Sodium chloride is a compound that consists of ions. There is no such thing as a molecule of sodium chloride. Instead, sodium chloride consists of sodium ions and chloride ions.

In many instances, the reaction between two atoms does not involve a complete loss and gain of electrons. Instead, electrons from both atoms are shared between the two atoms. In some cases, the sharing is equal, or nearly equal, with the electrons spending about half their time with each atom. In other cases, one atom will exert a somewhat stronger force on the electrons than the other atom. In that instance, the electrons are still shared by the two atoms—but not equally.

Electrons shared between two atoms are said to form a covalent bond. The combination of atoms joined to each other by means of a covalent bond is a molecule.

Polar and nonpolar molecules

Consider the situation when the electrons that make up a covalent bond spend more time with one atom than with the other. In that case, the atom that has the electrons more often will be slightly more negative than the other atom. The molecule that contains this arrangement is said to be a polar molecule. The term polar suggests a separation of charges, like the separation of magnetic force in a magnet with north and south poles.

But now think of a molecule in which the electrons in a covalent bond are shared equally—or almost equally. In that case, both atoms have the electrons about the same amount of time, and the distribution of negative electrical charge is about equal. There is no separation of charges, and the molecule is said to be nonpolar.

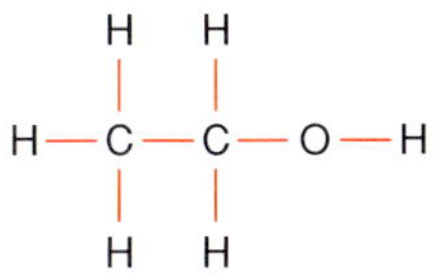

Figure 2. Structural formulas help differentiate between substances that share identical molecular formulas, such as ethyl alcohol and methyl ether. *(Reproduced by permission of The Gale Group.)*

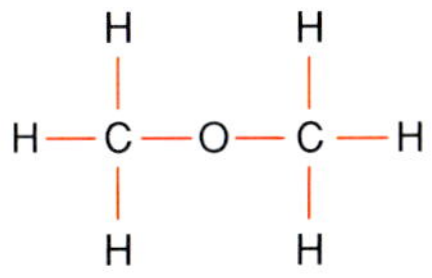

Ethyl alcohol
C_2H_6O

Methyl ether
C_2H_6O

Formulas

Molecular formulas. The structure of a molecule can be represented by a molecular formula. A molecular formula indicates the elements present in the molecule as well as the ratio of those elements. For example, the molecular formula for water is H_2O. That formula tells you, first of all, that two elements are present in the compound, hydrogen (H) and oxygen (O). The formula also tells that the ratio of hydrogen to oxygen in the compound is 2 to 1. (There is no 1 following the O in H_2O. If no number is written in as a subscript, it is understood to be 1.)

Structural formulas. A structural formula gives the same information as a molecular formula—the kind and number of atoms present—plus one more piece of information: the way those atoms are arranged within the molecule. As you'll notice in Figure 2, structural formulas help differentiate between substances that share identical molecular formulas, such as ethyl alcohol and methyl ether.

[*See also* **Atom; Chemical bond; Compound, chemical; Element, chemical; Formula, chemical**]

Mollusks

Mollusks belong to the phylum Mollusca and make up the second largest group of invertebrates (animals lacking backbones) after the arthropods. Over 100,000 species of mollusks have been identified. Restaurant menus often include a variety of mollusk dishes, such as oysters on the half-shell, steamed mussels, fried clams, fried squid, or escargots.

Mollusks have certain characteristic features, including a head with sense organs and a mouth, a muscular foot, a hump containing the digestive and reproductive organs, and an envelope of tissue (called the mantle) that usually secretes a hard, protective shell. Practically all of the shells found on beaches and prized by collectors belong to mollusks. Among the more familiar mollusks are snails, whelks, conchs, clams, mussels, scallops, oysters, squid, and octopuses. Less noticeable, but also common, are chitons, cuttlefish, limpets, nudibranchs, and slugs.

Classes of mollusks

The largest number of species of mollusks are in the class Gastropoda, which includes snails with a coiled shell and others lacking a shell. The next largest group are the bivalves (class Bivalvia), the chitons (class Amphineura), and octopus and squid, (class Cephalopoda). Other classes of mollusks are the class Scaphopoda, consisting of a few species of small mollusks with a tapered, tubular shell, and the class Monoplacophora. The last of these classes was once thought to be extinct, but a few living species have been found in the ocean depths. Some fossil shells recognizable as gastropods and bivalves have been found in rocks 570 million years old.

Evolutionary patterns

Mollusks provide a clear example of adaptive radiation. Adaptive radiation is the process by which closely related organisms gradually evolve in different directions in order to take advantage of specialized parts of the environment. The gastropods and bivalves were originally marine organisms, living in salt water. They subsequently evolved to take advantage of freshwater habitats. Without much change in their outward appearance, these animals developed physiological mechanisms to retain salts within their cells, a problem they did not face as marine organisms. This new development prevented excessive swelling of their bodies from intake of freshwater.

Several groups of freshwater snails then produced species adapted to life on land. The gills they originally used for the extraction of oxygen

A land snail. *(Reproduced by permission of JLM Visuals.)*

Momentum

The momentum of an object is defined as the mass of the object multiplied by the velocity of the object. Mathematically, that definition can be expressed as p = m · v, where p represents momentum, m represents mass, and v represents velocity.

In many instances, the mass of an object is measured in kilograms (kg) and the velocity in meters per second (m/s). In that case, momentum is measured in kilogram-meters per second (kg · m/s). Recall that velocity is a vector quantity. That is, the term velocity refers both to the speed with which an object is moving and to the direction in which it is moving. Since velocity is a vector quantity, then momentum must also be a vector quantity.

Conservation of momentum

Some of the most common situations involving momentum are those in which two moving objects collide with each other or in which a moving object collides with an object at rest. For example, what happens when two cars approach an intersection at the same time, do not stop, but collide with each other? In which direction will the cars be thrown, and how far will they travel after the collision?

The answer to that question can be obtained from the law of conservation of momentum, which says that the total momentum of a system before some given event must be the same as the total momentum of the system after the event. In this case, the total momentum of the two cars moving toward the intersection must be the same as the total momentum of the cars after the collision.

Suppose that the two cars are of very different sizes, a large Cadillac with a mass of 1,000 kilograms and a small Volkswagen with a mass of 500 kilograms, for example. If both cars are traveling at a velocity of 10 meters per second (mps), then the total momentum of the two cars is (for the Cadillac) 1,000 kg · 10 mps plus (for the Volkswagen) 500 kg · 10 mps = 10,000 kg · mps + 5,000 kg · mps = 15,000 kg · mps. Therefore, after the collision, the total momentum of the two cars must still be 15,000 kg · mps.

Applications

A knowledge of the laws of momentum is very important in many occupations. For example, the launch of a rocket provides a dramatic application of momentum conservation. Before launch, the rocket is at rest on the launch pad, so its momentum is zero. When the rocket engines fire, burning gases are expelled from the back of the rocket. By virtue of the law of conservation of momentum, the total momentum of the rocket and fuel must remain zero. The momentum of the escaping gases is regarded as having a negative value because they travel in a direction opposite to that of the rocket's intended motion. The rocket itself, then, must have momentum equal to that of the escaping gases, but in the opposite (positive) direction. As a result, the rocket moves forward.

[*See also* **Conservation laws; Mass; Laws of motion**]

Monsoon

A monsoon is a seasonal change in the direction of the prevailing wind. This wind shift typically brings about a marked change in local weather. Monsoons are often associated with rainy seasons in the tropics (the areas of Earth within 23.5 degrees latitude of the equator) and the subtropics (areas between 23.5 and about 35 degrees latitude, both north and south). In these areas, life is critically dependent on the monsoon rains. A weak monsoon rainy season may cause drought, crop failures, and hardship for people and wildlife. However, heavy monsoon rains have caused massive floods that have killed thousands of people.

Many parts of the world experience monsoons to some extent. Probably the most famous are the Asian monsoons, which affect India, China, Japan, and Southeast Asia. Monsoons also impact portions of central Africa, where their rain is critical to supporting life in the area south of the Sahara Desert. Lesser monsoon circulations affect parts of the southwestern United States. These summer rainy periods bring much needed rain to the dry plateaus of Arizona and New Mexico.

General monsoon circulation

Monsoons, like most other winds, occur in response to the Sun heating the atmosphere. In their simplest form, monsoons are caused by differences in temperatures between the oceans and continents. They are most likely to form where a large continental landmass meets a major ocean basin. During the early summer, the landmasses heat up more quickly than

Monsoon

> ## Words to Know
>
> **Circulation cell:** A circular path of air, in which warm air rises from the surface, moves to cooler areas, sinks back down to the surface, then moves back to near where it began. The air circulation sets up constant winds at the surface and aloft.
>
> **Convection:** The rising of warm air from the surface of Earth.
>
> **Jet stream:** High-speed winds that circulate around Earth at altitudes of 7 to 12 miles (12 to 20 kilometers) and affect weather patterns at the surface.
>
> **Subtropics:** Regions between 23.5 and about 35 degrees latitude, in both the northern and southern hemispheres, which surround the tropics.
>
> **Tropics:** Regions of Earth's surface lying within 23.5 degrees latitude of the equator.

ocean waters. The relatively warm land surface then heats the air over it, causing the air to convect, or rise. The convection of warm air produces an area of low pressure near the land surface. Meanwhile, air over the cooler ocean waters is humid, more dense, and under higher pressure.

The atmosphere always tries to maintain a balance by having air move into areas of low pressure from surrounding areas of high pressure. This movement is known as wind. Thus during the summer, oceanic air flows toward the low pressure over land. This flow is continually supplied by cooler oceanic air sinking from higher levels in the atmosphere. In the upper atmosphere, the rising continental (landmass) air is drawn outward over the oceans to replace the sinking oceanic air, thus completing the cycle. In this way a large vertical circulation cell is set up, driven by solar heating. At the surface, the result is a constant wind flowing from sea to land.

As it flows onto shore, the moist ocean air is pulled upward as part of the convecting half of the circulation cell. The rising air cools and soon can no longer contain moisture. Eventually rain clouds form. Rain clouds are especially likely to occur when the continental areas have higher elevations (mountains, plateaus, etc.) because the humid ocean air is forced upward over these barriers, causing widespread cloud formation and heavy rains. This is the reason why the summer monsoon forms the rainy season in many tropical areas.

In the late fall and early winter, the situation is reversed. Land surfaces cool off quickly in response to cooler weather, but the same property of water that makes it slow to absorb heat also causes it to cool slowly. As a result, continents are usually cooler than the oceans surrounding them during the winter. This sets up a new circulation in the reverse direction: air over the sea, now warmer than that over the land, rises and is replaced by winds flowing off the continent. The continental winds are supplied by cooler air sinking from aloft. At upper atmospheric levels, the rising oceanic air moves over the land to replace the sinking continental air. Sinking air (high pressure) prevents the development of clouds and rain, so during the winter monsoon continental areas are typically very dry. This winter circulation causes a prevailing land-to-sea wind until it collapses with the coming of spring.

The monsoon of India

The world's most dramatic monsoon occurs in India. During the early summer months, increased solar heating begins to heat the Indian subcontinent, which would tend to set up a monsoon circulation cell between southern Asia and the Indian Ocean. However, the development of the summer monsoon is delayed by the subtropical jet stream.

Jet streams are great rivers of air that ring Earth at levels in the atmosphere ranging from 7 to 8 miles (11 to 13 kilometers) above the surface. The subtropical jet stream is a permanent feature, flowing westerly (from west to east). It migrates over the year in response to the seasons, moving northward to higher latitudes in the summer and southward in the winter.

As summer progresses, the subtropical jet slides northward. The extremely high Himalayan mountains present an obstacle for the jet; it must "jump over" the mountains and reform over central Asia. When it finally does so, a summer monsoon cell develops. The transition can be very fast: the Indian monsoon has a reputation for appearing suddenly as soon as the subtropical jet stream is out of the way. As the air is forced to rise over the foothills of the Himalayas, it causes constant, heavy rains, often resulting in destructive flooding. The town of Cherrapunji, India, located on the Himalayan slopes, receives an annual rainfall of over 36 feet (11 meters), making it one of the wettest places on Earth.

When the monsoon fails

The importance of monsoons is demonstrated by the experience of the Sahel, a band of land on the southern fringe of Africa's Sahara Desert.

Moon

The rains of the seasonal monsoon normally transform this arid (dry) area to a grassland suitable for grazing livestock. The wetter southern Sahel can support farming, and many residents migrated to the area during the years of strong monsoons. Beginning in the late 1960s, however, the annual monsoons began to fail. The pasture areas in the northern Sahel dried up, forcing wandering herders and their livestock southward in search of pasture and water. The monsoon rains did not return until 1974. In the intervening six years, the area suffered devastating famines and loss of life, both human and animal.

[*See also* **Atmospheric circulation; El Niño**]

Moon

The Moon is a roughly spherical, rocky body orbiting Earth at an average distance of 240,00 miles (385,000 kilometers). It measures about 2,160 miles (3,475 kilometers) across, a little over one-quarter of Earth's diameter. Earth and the Moon are the closest in size of any known planet and its satellite, with the possible exception of Pluto and its moon Charon.

The Moon is covered with rocks, boulders, craters, and a layer of charcoal-colored soil from 5 to 20 feet (1.5 to 6 meters) deep. The soil consists of rock fragments, pulverized rock, and tiny pieces of glass. Two types of rock are found on the Moon: basalt, which is hardened lava; and breccia, which is soil and rock fragments that have melted together.

Elements found in Moon rocks include aluminum, calcium, iron, magnesium, titanium, potassium, and phosphorus. In contrast with Earth, which has a core rich in iron and other metals, the Moon appears to contain very little metal. The apparent lack of organic compounds rules out the possibility that there is, or ever was, life on the Moon.

The Moon has no weather, no wind or rain, and no air. As a result, it has no protection from the Sun's rays or meteorites and no ability to retain heat. Temperatures on the Moon have been recorded in the range of 280°F (138°C) to −148°F (−100°C).

Formation of the Moon

Both Earth and the Moon are about 4.6 billion years old, a fact that has led to many theories about their common origin. Before the 1970s, scientists held to one of three competing theories about the origin of the Moon: the fission theory, the simultaneous creation theory, and the capture theory.

The fission theory stated that the Moon spun off from Earth early in its history. The Pacific basin was the scar left by the tearing away of the Moon. The simultaneous creation theory stated that the Moon and Earth formed at the same time from the same planetary building blocks that were floating in space billions of years ago. The capture theory stated that the Moon was created somewhere else in the solar system and captured by Earth's gravitational field as it wandered too close to the planet.

After scientists examined the age and composition of lunar rocks brought back by Apollo astronauts, they discarded these previous theories and accepted a new one: the giant impact theory (also called the Big Whack model). This theory states that when Earth was newly formed, it was sideswiped by a celestial object that was at least as massive as Mars. (Some scientists contend the object was two to three times the mass of Mars.) The collision spewed a ring of crustal matter into space. While in orbit around Earth, that matter gradually combined to form the Moon.

A photo of the full moon taken from *Apollo 17*. The flatter regions—called mares—appear as dark areas because they reflect less light. The highlands are lighter in color and have a more rugged surface. *(Reproduced by permission of JLM Visuals.)*

Moon

The evolution of the Moon has been completely different from that of Earth. For about the first 700 million years of the Moon's existence, it was struck by great numbers of meteorites. They blasted out craters of all sizes. The sheer impact of so many meteorites caused the Moon's crust to melt. Eventually, as the crust cooled, lava from the interior surfaced and filled in cracks and some crater basins. These filled-in basins are the dark spots we see when we look at the Moon.

To early astronomers, these dark regions appeared to be bodies of liquid. In 1609, Italian astronomer Galileo Galilei became the first person to observe the Moon through a telescope. He named these dark patches "maria," Latin for "seas."

In 1645, Polish astronomer Johannes Hevelius, known as the father of lunar topography, charted 250 craters and other formations on the Moon. Many of these were later named for philosophers and scientists, such as Danish astronomer Tycho Brahe, Polish astronomer Nicolaus Copernicus, German astronomer Johannes Kepler, and Greek philosopher Plato.

Humans on the Moon

All Earth-based study of the Moon has been limited by one factor: only one side of the Moon ever faces Earth. The reason is that the Moon's rotational period is equal to the time it takes the Moon to complete one orbit around Earth. It wasn't until 1959, when the former Soviet Union's space probe *Luna 3* traveled to the far side of the Moon that scientists were able to see the other half for the first time.

Then in 1966, the Soviet *Luna 9* became the first object from Earth to land on the Moon. It took television footage showing that lunar dust, which scientists had anticipated finding, did not exist. The fear of encountering thick layers of dust was one reason both the Soviet Union and the United States hesitated sending a man to the moon.

Just three years later, on July 20, 1969, U.S. astronauts Neil Armstrong and Edwin "Buzz" Aldrin aboard *Apollo 11* became the first humans to walk on the Moon. They collected rock and soil samples, from which scientists learned the Moon's elemental composition. There were five more lunar landings in the Apollo program between 1969 and 1972. To this day, the Moon remains the only celestial body to be visited by humans.

Water on the Moon?

In late 1996, scientists announced the possibility that water ice existed on the Moon. *Clementine,* a U.S. Defense Department spacecraft, had been launched in January 1994 and orbited the Moon for four months. It surveyed a huge depression in the south polar region called the South Pole-Aitken basin. Nearly four billion years ago, a massive asteroid had gouged out the basin. It stretches 1,500 miles (2,415 kilometers) and in places is as deep as 8 miles (13 kilometers), deeper than Mount Everest is high.

Areas of this basin are never exposed to sunlight, and temperatures there are estimated to be as low as −387°F (−233°C). While scanning these vast areas with radar signals, *Clementine* discovered what appeared to be ice crystals mixed with dirt. Scientists speculated that the crystals made up no more the 10 percent of the material in the region. They believe the ice is the residue of moisture from comets that struck the Moon over the last three billion years.

To learn more about the Moon and this possible ice, the National Aeronautics and Space Administration (NASA) launched the *Lunar Prospector* in January 1998. This was NASA's first mission back to the Moon in 25 years. As the name of this small, unmanned spacecraft implied, its nineteen-month mission was to "prospect" the surface composition of the Moon, providing a detailed map of minerals, water ice, and certain gases. It also took measurements of magnetic and gravity fields, and tried to provide scientists with information regarding the size and content of the Moon's core. For almost a year, *Lunar Prospector* orbited the Moon at an altitude

The first footprint on the moon. *(Reproduced by permission of National Aeronautics and Space Administration.)*

of 62 miles (100 kilometers). Then, in December 1998, NASA lowered its orbit to an altitude of 25 miles (40 kilometers). On July 31, 1999, in a controlled crash, the spacecraft settled into a crater near the south pole of the Moon. If there were water at the crash site, the spacecraft's impact would have thrown up a huge plume of water vapor that could have been seen by spectroscopes at the Keck Observatory on Mauna Kea, Hawaii, and other telescopes like the orbiting Hubble Space Telescope. However, no such plume was observed. For scientists, the question of whether there is hidden ice on the Moon, delivered by impacting comets, is still open. It is estimated that each pole on the Moon may contain up to 1 billion tons (900 million metric tons) of frozen water ice spread throughout the soil.

[See also **Orbit; Satellite; Spacecraft, manned**]

Mounds, earthen

Earthen mounds are raised banks or hills built by prehistoric humans almost entirely out of soil or earth. Found in many different parts of the world, these mounds vary in size and shape, and most were built by ancient peoples as burial places or to serve some ceremonial purpose. The greatest number and the most famous earthen mounds were built by early Native Americans.

Mounds are made by people

An earthen mound is an above-ground pile of earth that often looks like a large, rounded bump on Earth's surface or sometimes more like a normal, natural hill. Mounds still exist in many parts of the world and were usually built by humans long ago to bury their dead. Different countries and cultures call them by different names, and they range in size from a few feet or meters across to huge, pyramidlike structures that contain tons of earth. Although the earthen mounds found today in North America are similar to those discovered in Europe and Asia, these American mounds are so numerous and varied that the name "mound builders" has come to refer to those early Native Americans who constructed large monuments out of earth.

Different mounds for different purposes

Tens of thousands of earthen mounds can still be seen from the Canadian provinces of Ontario and Manitoba south to Florida, and from the Atlantic Ocean to the Mississippi River. They were built by several different groups of Native American people who may have lived as long ago as 1000 B.C. While these mounds take many forms and served different purposes, with each in a sense telling its own story, all were built entirely by hand, usually by piling up earth one basket-full at a time. Some served as burial mounds for the honored dead, while other flat-topped mounds were parts of large cities or towns and held temples or ceremonial buildings, and still others were built in the shape of giant animals. These huge, raised mounds are easily recognizable from the air and resemble the outline of a certain animal, like a snake, bird, or bear. They are called "effigy mounds" (pronounced EFF-ih-jee). Today we realize that the mound builders were not a single group of people, and that their mounds were not built only one way for a single purpose.

Early investigators

When Spanish explorer Hernando de Soto (c.1496–1542) landed in Florida in 1539 and traveled southwest, he wrote of noticing that each

Mounds, earthen

The serpent mounds of southern Ohio. *(Reproduced by permission of Photo Researchers, Inc.)*

Mounds, earthen

native town he encountered had one or more of these high, artificial mounds. Over 200 years later, one of the first people to investigate these American mounds was Thomas Jefferson (1743–1826), who went on to become third president of the United States. Sometime around 1780, when Jefferson was governor of Virginia, he excavated or dug up and exposed some of the burial mounds in Virginia. Digging carefully, Jefferson approached this job as a modern archaeologist would, and although he uncovered many human skeletons, he was not searching just for buried treasure or ancient goods. Despite his belief that these mounds were the work of Native Americans, a myth soon grew up that they were instead built by some sophisticated lost people who had lived long before. This wrong notion persisted for quite some time, until it was finally disproved during the 1880s by surveying and excavating teams sponsored by the Smithsonian Institution. Their work eventually demonstrated that these mysterious mounds were the work of ancient Native Americans, and the mounds eventually came to be protected by state laws.

Adena burial mounds

One of the earliest groups of Native American mound builders was located in the Ohio River valley. Today, the people of this group are known as the people of the Adena culture. These people probably believed in some sort of afterlife because they conducted burial ceremonies and built mounds for their dead. Many of these began as single heaps of earth covered by simple monuments of stone and other materials. As bodies were later added to a mound, it grew in size, and sometimes special earth-covered log tombs were built to contain high-ranking tribe members. Often they would be buried with objects such as pipes, pottery, axes, and other gifts. One of the largest Adena mounds, measuring about 70 feet (21 meters) high, is in West Virginia.

Hopewell mounds

The Adena people were succeeded by the Hopewell culture in what is now Michigan, Wisconsin, Ohio, Indiana, Iowa, and Missouri. This group is named after a farm in Ohio where about 30 mounds are located. The people of the Hopewell culture traveled and traded as far away as Florida, bringing back shark teeth and seashells to bury with their dead. They built more mounds than the Adena people, and the largest, in Newark, Ohio, includes a raised ridge that surrounds about 50 acres (20 hectares) of land. Their mounds almost always contained gifts for their dead.

Mississippian culture

The Hopewell culture eventually declined for some reason, and they were succeeded by what we call the Mississippians, because these people made their mounds in the Mississippi valley. They were naturally more advanced, and built actual cities with many flat-topped pyramids. Their mounds served as foundations for temples or special buildings as well as for burial places. It is thought that they adopted many of the customs they encountered during their trade visits to Mexico. One of the largest mound sites in the United States is Poverty Point, near Epps, Louisiana. It may be 3,000 years old and probably served as a ceremonial center for the culture of the time. It consists of a group of six octagons (eight-sided shapes), spreading out one within the other, with the outer octagon having a diameter of about 4,000 feet (1,220 meters).

Importance of mounds

The existence of these mounds tell us something about the people who built them, especially when they contain objects. Study and understanding of the mounds can tell us something about that group's society, or how they lived and what they were like. Most important, the mounds are proof that advanced cultures existed in ancient America long before the Europeans came. We now know that we should recognize and respect these cultures, preserving and protecting what they have left behind.

Mountain

A mountain is any landmass on Earth's surface that rises to a great height in comparison to its surrounding landscape. Mountains usually have more-or-less steep sides meeting in a summit that is much narrower in width than the mountain's base.

Although single mountains exist, most occur as a group, called a mountain range. A group of ranges that share a common origin and form is known as a mountain system. A group of systems is called a mountain chain. Finally, a complex group of continental (land-based) ranges, systems, and chains is called a mountain belt or cordillera (pronounced kor-dee-YARE-ah).

The greatest mountain systems are the Alps of Europe, the Andes of South America, the Himalayas of Asia, and the Rockies of North America. Notable single peaks in these systems include Mont Blanc (Alps), Aconcagua (Andes), Everest (Himalayas), and Elbert (Rockies). The

Mountain

> **Words to Know**
>
> **Belt:** Complex group of continental mountain ranges, systems, and chains.
>
> **Chain:** Group of mountain systems.
>
> **Crust:** Thin layer of rock covering the planet.
>
> **Lithosphere:** Rigid uppermost section of the mantle combined with the crust.
>
> **Orogeny:** Mountain building.
>
> **Plate tectonics:** Geological theory holding that Earth's surface is composed of rigid plates or sections that move about the surface in response to internal pressure, creating the major geographical features such as mountains.
>
> **Range:** Group of mountains.
>
> **System:** Group of mountain ranges that share a common origin and form.

Himalayas is the world's highest mountain system, containing some 30 peaks rising to more than 25,000 feet (7,620 meters). Included among these peaks is the world's highest, Mount Everest, at 29,028 feet (8,848 meters) above sea level. North America's highest peak is Mount McKinley, part of the Alaska Range, which rises 20,320 feet (6,194 meters).

Mountains, like every other thing in the natural world, go through a life cycle. They rise from a variety of causes and wear down over time at various rates. Individual mountains do not last very long in the powerfully erosive atmosphere of Earth. Mountains on the waterless world of Mars are billions of years old, but Earth's peaks begin to fracture and dissolve as soon as their rocks are exposed to the weathering action of wind and rain. This is why young mountains are high and rugged, while older mountains are lower and smoother.

Mountain building

Mountain building (a process known as orogeny [pronounced o-RA-je-nee]) occurs mainly as a result of movements in the surface of Earth. The thin shell of rock covering the globe is called the crust, which varies in depth from 5 to 25 miles (8 to 40 kilometers). Underneath the crust is the mantle, which extends to a depth of about 1,800 miles (2,900 kilometers) below the surface. The mantle has an upper rigid layer and a partially melted lower layer. The crust and the upper rigid layer of the mantle together make up the lithosphere. The lithosphere, broken up into various-sized plates or sections, "floats" on top of the heated, semiliquid layer underneath.

The heat energy carried from the core of the planet through the semiliquid layer of the mantle causes the lithospheric plates to move back and forth. This motion is known as plate tectonics. Plates that move toward each other are called convergent plates; plates moving away from each other are divergent plates.

When continental plates converge, they shatter, fold, and compress the rocks of the collision area, thrusting the pieces up into a mountain range of great height. This is how the Appalachians, Alps, and Himalayas were formed: the rocks of their continents were folded just as a flat-lying piece of cloth folds when pushed.

When a continental plate and an oceanic plate converge, the oceanic plate subducts or sinks below the continental plate because it is more dense. As the oceanic plate sinks deeper and deeper into Earth, its leading edge of rock is melted by intense pressure and heat. The molten rock then rises to the surface where it lifts and deforms rock, resulting in the formation of volcanic mountains on the forward edge of the continental plate. The Andes and the Cascade Range in the western United States are examples of this type of plate convergence.

The longest mountain range on Earth is entirely underwater. The Mid-Atlantic Ridge is a submarine mountain range that extends about 10,000 miles (16,000 kilometers) from Iceland to near the Antarctic Circle. The ridge is formed by the divergence of two oceanic plates. As the plates move away from each other, magma (molten rock) from inside Earth rises and creates new ocean floor in a deep crevice known as a rift valley in the middle of the ridge. On either side of the rift lie tall volcanic mountains. The peaks of some of these mountains rise above the surface of the ocean to form islands, such as Iceland and the Azores.

Other mountains on the planet form as solitary volcanic mountains in rift valleys on land where two continental plates are diverging. Mount Kilimanjaro, the highest point in Africa, is an extinct volcano that stands along the Great Rift Valley in northeast Tanzania. The highest of its two peaks, Kibo, rises 19,340 feet (5,895 meters) above sea level.

The erosive power of water on plateaus can also create mountains. Mesas, flat-topped mountains common in the southwest United States, are

Mountain

such a case. They form when a solid sheet of hard rock sits on top of softer rock. The hard rock layer on top, called the caprock, once covered a wide area. The caprock is cut up by the erosive action of streams. Where there is no more caprock, the softer rock beneath washes away relatively quickly. Mesas are left wherever a remnant of the caprock forms a roof over the softer rock below. Mesa Verde in Colorado and the Enchanted Mesa in New Mexico are classic examples.

Mountains and weather

Mountains make a barrier for moving air, robbing it of any precipitation. The atmosphere at higher elevations is cooler and thinner. As dense masses of warm, moist air are pushed up a mountain slope by winds, the air pressure surrounding the mass drops away. As a result, the mass becomes cooler. The moisture contained in the mass then condenses into cool droplets, and clouds form over the mountain. As the clouds continue to rise into cooler, thinner air, the droplets increase in size until they become too heavy to float in the air. The clouds then dump rain or snow on the mountain slope. After topping the crest, however, the clouds often contain little moisture to rain on the lee side of the mountain, which becomes arid. This is best illustrated in the Sierra Nevada mountains of

Because mountains work as a barrier for moving air, they are often topped with snow caused by the cold precipitation in the clouds surrounding the peaks. (Reproduced by permission of JLM Visuals.)

1 3 0 4   U·X·L Encyclopedia of Science, 2nd Edition

California, where tall redwood forests cover the ocean-facing side of the mountains and Death Valley lies on the lee side.

[*See also* **Plate tectonics; Volcano**]

Multiple personality disorder

Multiple personality disorder (MPD) is a chronic (recurring frequently) emotional illness. A person with MPD plays host to two or more personalities (called alters). Each alter has its own unique style of viewing and understanding the world and may have its own name. These distinct personalities periodically control that person's behavior as if several people were alternately sharing the same body.

MPD occurs about eight times more frequently in women than in men. Some researchers believe that because men with MPD tend to act more violently than women, they are jailed rather than hospitalized and, thus, never diagnosed. Female MPD patients often have more identities than men, averaging fifteen as opposed to eight for males.

Causes of multiple personality disorder

Most people diagnosed with MPD were either physically or sexually abused as children. Many times when a young child is severely abused, he or she becomes so detached from reality that what is happening may seem more like a movie or television show than real life. This self-hypnotic state, called disassociation, is a defense mechanism that protects the child from feeling overwhelmingly intense emotions. Disassociation blocks off these thoughts and emotions so that the child is unaware of them. In effect, they become secrets, even from the child. According to the American Psychiatric Association, many MPD patients cannot remember much of their childhoods.

Not all children who are severely and repeatedly abused develop multiple personality disorder. However, if the abuse is repeatedly extreme and the child does not have enough time to recover emotionally, the disassociated thoughts and feelings may begin to take on lives of their own. Each cluster of thoughts tends to have a common emotional theme such as anger, sadness, or fear. Eventually, these clusters develop into full-blown personalities, each with its own memory and characteristics.

Multiple personality disorder

> ### Words to Know
>
> **Alter:** Alternate personality that has split off or disassociated from the main personality, usually after severe childhood trauma.
>
> **Disassociation:** Separation of a thought process or emotion from conscious awareness.
>
> **Hypnosis:** Trance state during which people are highly vulnerable to the suggestions of others.
>
> **Personality:** Group of characteristics that motivates behavior and sets us apart from other individuals.
>
> **Switching:** Process by which an alternate personality reveals itself and controls behavior.
>
> **Trauma:** An extremely severe emotional shock.

Symptoms of the disorder

A person diagnosed with MPD can have as many as a hundred or as few as two separate personalities. (About half of the recently reported cases have ten or fewer.) These different identities can resemble the normal personality of the person or they may take on that of a different age, sex, or race. Each alter can have its own posture, set of gestures, and hairstyle, as well as a distinct way of dressing and talking. Some may speak in foreign languages or with an accent. Sometimes alters are not human, but are animals or imaginary creatures.

The process by which one of these personalities reveals itself and controls behavior is called switching. Most of the time the change is sudden and takes only seconds. Sometimes it can take hours or days. Switching is often triggered by something that happens in the patient's environment, but personalities can also come out under hypnosis (a trancelike state in which a person becomes very responsive to suggestions of others).

Sometimes the most powerful alter serves as the gatekeeper and tells the weaker alters when they may reveal themselves. Other times alters fight each other for control. Most patients with MPD experience long periods during which their normal personality, called the main or core personality, remains in charge. During these times, their lives may appear normal.

Ninety-eight percent of people with MPD have some degree of amnesia when an alter surfaces. When the main personality takes charge once again, the time spent under control of an alter is completely lost to memory. In a few instances, the host personality may remember confusing bits and pieces of the past. In some cases alters are aware of each other, while in others they are not.

One of the most baffling mysteries of MPD is how alters can sometimes show very different biological characteristics from the host and from each other. Several personalities sharing one body may have different heart rates, blood pressures, body temperatures, pain tolerances, and eyesight abilities. Different alters may have different reactions to medications. Sometimes a healthy host can have alters with allergies and even asthma.

Treatment

MPD does not disappear without treatment, although the rate of switching seems to slow down in middle age. The most common treatment for MPD is long-term psychotherapy twice a week. During these sessions, the therapist must develop a trusting relationship with the main personality and each of the alters. Once that is established, the emotional issues of each personality regarding the original trauma are addressed. The main and alters are encouraged to communicate with each other in order to integrate or come together. Hypnosis is often a useful tool to accomplish this goal. At the same time, the therapist helps the patient to acknowledge and accept the physical or sexual abuse he or she endured as a child and to learn new coping skills so that disassociation is no longer necessary.

About one-half of all people being treated for MPD require brief hospitalization, and only 5 percent are primarily treated in psychiatric hospitals. Sometimes mood-altering medications such as tranquilizers or antidepressants are prescribed for MPD patients. The treatment of MPD lasts an average of four years.

Multiplication

Multiplication is often described as repeated addition. For example, the product 3×4 is equal to the sum of three 4s: $4 + 4 + 4$.

Terminology

In talking about multiplication, several terms are used. In the expression 3×4, the entire expression, whether it is written as 3×4 or as

Multiplication

> ### Words to Know
>
> **Factor:** A number used as a multiplier in a product.
>
> **Multiplier:** One of two or more numbers combined by multiplication to form a product.
>
> **Product:** The result of multiplying two or more numbers.

12, is called the product. In other words, the answer to a multiplication problem is the product. In the original expression, the numbers 3 and 4 are each called multipliers, factors, or terms. At one time, the words multiplicand and multiplier were used to indicate which number got multiplied (the multiplicand) and which number did the multiplying (the multiplier). That terminology has now fallen into disuse. Now the term multiplier applies to either number.

Multiplication is symbolized in three ways: with an ×, as in 3 × 4; with a centered dot, as in 3 · 4; and by writing the numbers next to each other, as in 3(4), (3)(4), 5x, or (x + y)(x − y).

Rules of multiplication for numbers other than whole—or natural—numbers

Common fractions. The numerator of the product is the product of the numerators; the denominator of the product is the product of the denominators. For example, $(\frac{3}{7})(\frac{5}{4}) = \frac{15}{28}$.

Decimals. Multiply the decimal fractions as if they were natural numbers. Place the decimal point in the product so that the number of places in the product is the sum of the number of places in the multipliers. For example, 3.07 × 5.2 = 15.964.

Signed numbers. Multiply the numbers as if they had no signs. If the two factors both have the same sign, give the product a positive sign or omit the sign entirely. If the two factors have different signs, give the product a negative sign. For example, (3x)(−2y) = −6xy; (−5)(−4) = +20.

Powers of the same base. To multiply two powers of the same base, add the exponents. For example $10^2 \times 10^3 = 10^5$ and $x^5 \times x^{-2} = x^3$.

Monomials. To multiply two monomials, find the product of the numerical and literal parts of the factors separately. For example, $(3x^2y)(5xyz) = 15x^3y^2z$.

Polynomials. To multiply two polynomials, multiply each term of one by each term of the other, combining like terms. For example, $(x + y)(x - y) = x^2 - xy + xy - y^2 = x^2 - y^2$.

Applications

Multiplication is used in almost every aspect of our daily lives. Suppose you want to buy three cartons of eggs, each containing a dozen eggs, at 79 cents per carton. You can find the total number of eggs purchased (3 cartons times 12 eggs per carton = 36 eggs) and the cost of the purchase (3 cartons at 79 cents per carton = $2.37).

Specialized professions use multiplication in an endless variety of ways. For example, calculating the speed with which the Space Shuttle will lift off its launch pad involves untold numbers of multiplication calculations.

Muscular system

The muscular system is the body's network of tissues that controls movement both of the body and within it (such as the heart's pumping action and the movement of food through the gut). Movement is generated through the contraction and relaxation of specific muscles.

The muscles of the body are divided into two main classes: skeletal (voluntary) and smooth (involuntary). Skeletal muscles are attached to the skeleton and move various parts of the body. They are called voluntary because a person controls their use, such as in the flexing of an arm or the raising of a foot. There are about 650 skeletal muscles in the whole human body. Smooth muscles are found in the stomach and intestinal walls, vein and artery walls, and in various internal organs. They are called involuntary muscles because a person generally cannot consciously control them. They are regulated by the autonomic nervous system (part of the nervous system that affects internal organs).

Another difference between skeletal and smooth muscles is that skeletal muscles are made of tissue fibers that are striated or striped. These alternating bands of light and dark result from the pattern of the filaments (threads) within each muscle cell. Smooth muscle fibers are not striated.

Muscular system

> ### Words to Know
>
> **Autonomic nervous system:** Part of the nervous system that regulates involuntary action, such as of the heart and intestines.
>
> **Extensor muscle:** Muscle that contracts and causes a joint to open.
>
> **Flexor muscle:** Muscle that contracts and causes a joint to close.
>
> **Myoneural juncture:** Area where a muscle and a nerve connect.
>
> **Tendon:** Tough, fibrous connective tissue that attaches muscle to bone.

The cardiac or heart muscle (also called myocardium) is a unique type of muscle that does not fit clearly into either of the two classes of muscle. Like skeletal muscles, cardiac muscles are striated. But like smooth muscles, they are involuntary, controlled by the autonomic nervous system.

The longest muscle in the human body is the sartorius (pronounced sar-TOR-ee-us). It runs from the waist down across the front of thigh to the knee. Its purpose is to flex the hip and knee. The largest muscle in the body is the gluteus maximus (pronounced GLUE-tee-us MAX-si-mus; buttocks muscles). It moves the thighbone away from the body and straightens out the hip joint.

Skeletal muscles

Skeletal muscles are probably the most familiar type of muscle. They are the muscles that ache after strenuous work or exercise. Skeletal muscles make up about 40 percent of the body's mass or weight. They stabilize joints, help maintain posture, and give the body its general shape. They also use a great deal of oxygen and nutrients from the blood supply.

Skeletal muscles are attached to bones by tough, fibrous connective tissue called tendons. Tendons are rich in the protein collagen, which is arranged in a wavy way so that it can stretch and provide additional length at the muscle-bone junction.

Skeletal muscles act in pairs. The flexing (contracting) of one muscle is balanced by a lengthening (relaxation) of its paired muscle or a

Muscular system

group of muscles. These antagonistic (opposite) muscles can open and close joints such as the elbow or knee. An example of antagonistic muscles are the biceps (muscles in the front of the upper arm) and the triceps (muscles in the back of the upper arm). When the biceps muscle flexes, the forearm bends in at the elbow toward the biceps; at the same time, the triceps muscle lengthens. When the forearm is bent back out in a straight-arm position, the biceps lengthens and the triceps flexes.

Muscles that contract and cause a joint to close, such as the biceps, are called flexor muscles. Those that contract and cause a joint to open, such as the triceps, are called extensors. Skeletal muscles that support the skull, backbone, and rib cage are called axial skeletal muscles. Skeletal muscles of the limbs (arms and legs) are called distal skeletal muscles.

Skeletal muscle fibers are stimulated to contract by electrical impulses from the nervous system. Nerves extend outward from the spinal cord to connect to muscle cells. The area where a muscle and a nerve connect is called the myoneural juncture. When instructed to do so, the nerve releases a chemical called a neurotransmitter that crosses the microscopic space between the nerve and the muscle and causes the muscle to contract.

Skeletal muscle fibers are characterized as fast or slow based on their activity patterns. Fast (also called white) muscle fibers contract

Close-up of striated skeletal muscle. *(Reproduced by permission of Photo Researchers, Inc.)*

Muscular system

rapidly, have poor blood supply, operate without oxygen, and tire quickly. Slow (also called red) muscle fibers contract more slowly, have better blood supplies, operate with oxygen, and do not tire as easily. Slow muscle fibers are used in movements that are ongoing, such as maintaining posture.

Smooth muscles

Smooth muscle fibers line most of the internal hollow organs of the body, such as the intestines, stomach, and uterus (womb). They help move substances through tubular areas such as blood vessels and the small intestines. Smooth muscles contract automatically, spontaneously, and often rhythmically. They are slower to contract than skeletal muscles, but they can remain contracted longer.

Like skeletal muscles, smooth muscles contract in response to neurotransmitters released by nerves. Unlike skeletal muscles, some smooth muscles contract after being stimulated by hormones (chemicals secreted by glands). An example is oxytocin, a hormone released by the pituitary gland. It stimulates the smooth muscles of the uterus to contract during childbirth.

Smooth muscles are not as dependent on oxygen as skeletal muscles are. Smooth muscles use carbohydrates to generate much of their energy.

Human skeletal muscles (anterior view). *(Reproduced by permission of Photo Researchers, Inc.)*

Cardiac muscle

The cardiac muscle or myocardium contracts (beats) more than 2.5 billion times in an average lifetime. Like skeletal muscles, myocardium is striated. However, myocardial muscle fibers are smaller and shorter than skeletal muscle fibers.

The contractions of the myocardium are stimulated by an impulse sent out from a small clump (node) of specialized tissue in the upper right area of the heart. The impulse spreads across the upper area of the heart, causing this region to contract. This impulse also reaches another node, located near the lower right area of the heart. After receiving the initial impulse, the second node fires off its own impulse, causing the lower region of the heart to contract slightly after the upper region.

Disorders of the muscular system

The most common muscular disorder is injury from misuse. Skeletal muscle sprains and tears cause excess blood to seep into the tissue in order to heal it. The remaining scar tissue results in a slightly shorter muscle. Overexertion or a diminished blood supply can cause muscle cramping. Diminished blood supply and oxygen to the heart muscle causes chest pain called angina pectoris.

The most common type of genetic (inherited) muscular disorder is muscular dystrophy. This disease causes muscles to progressively waste away. There are six forms of muscular dystrophy. The most frequent and most dreaded form appears in boys aged three to seven. (Boys are usually affected because it is a sex-linked condition; girls are carriers of the disease and are usually not affected.) The first symptom of the disease is a clumsiness in walking. This occurs because the muscles of the pelvis and the thighs are first affected. The disease spreads to muscles in other areas of the body, and by the age of ten, a child is usually confined to a wheelchair or a bed. Death usually occurs before adulthood.

Another form of muscular dystrophy appears later in life and affects both sexes equally. The first signs of the disease appear in adolescence. The muscles affected are those in the face, shoulders, and upper arms. People with this form of the disease may survive until middle age.

Currently, there is no known treatment or cure for any form of muscular dystrophy.

[*See also* **Heart**]

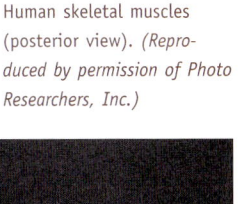

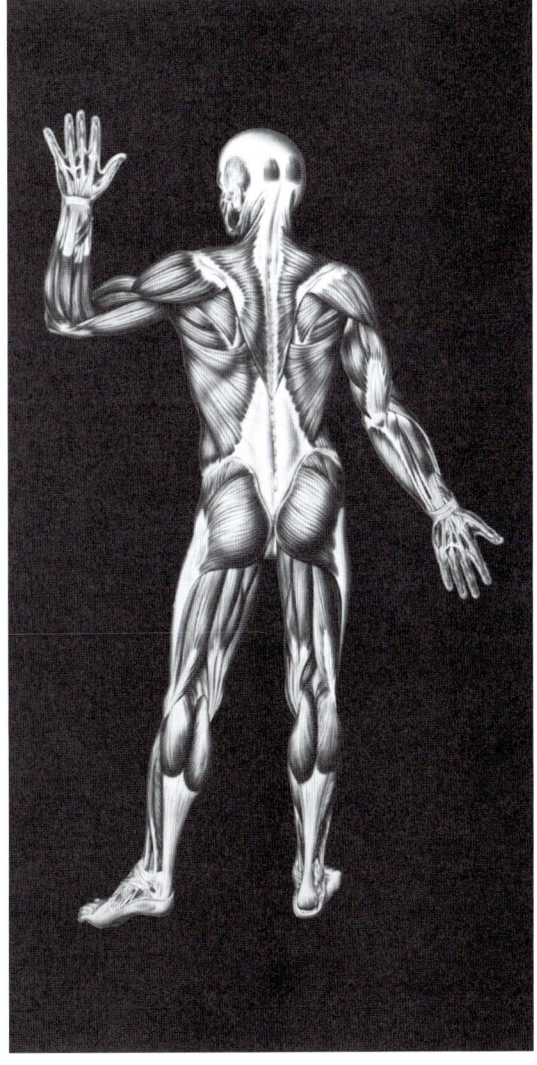

Human skeletal muscles (posterior view). *(Reproduced by permission of Photo Researchers, Inc.)*

Mutation

A mutation is a permanent change in a gene that is passed from one generation to the next. An organism born with a mutation can look very different from its parents. People with albinism—the lack of color in the skin, hair, and eyes—have a mutation that eliminates skin pigment. Dwarfs are an example of a mutation that affects growth hormones.

Mutations are usually harmful and often result in the death of an organism. However, some mutations may help an organism survive or be beneficial to a species as a whole. In fact, useful mutations are the driving force behind evolution.

Changes in DNA

Until the mid-1950s, no explanation for the sudden appearance of mutations existed. Today we know that mutations are caused when the hereditary material of life is altered. That hereditary material consists of long, complex molecules known as deoxyribonucleic acid (DNA).

Every cell contains DNA on threadlike structures called chromosomes. Sections of a DNA molecule that are coded to create specific proteins are known as genes. Proteins are chemicals produced by the body that are vital to cell function and structure. Human beings carry about 100,000 genes on their chromosomes. If the structure of a particular gene is altered, that gene will no longer be able to perform the function it is supposed to perform. The protein for which it codes will also be missing or defective. Just one missing or abnormal protein can have a dramatic effect on the entire body. Albinism, for instance, is caused by the loss of one single protein.

A molecule of DNA itself is made up of subunits known as nucleotides. Four different nucleotides are used in DNA molecules. They are commonly abbreviated by the letters A, C, G, and T. A typical DNA molecule could be represented, for example, as shown below:

-A-T-C-T-C-T-G-G-C-C-C-A-G-T-C-C-G-T-T-G-A-T-G-C-T-G-T-

Each group of three nucleotides means something specific to a cell. For example, the nucleotide CCT tells a cell to make the amino acid glycine. The string of nucleotides shown above, when read three at a time, then, tells a cell which amino acids to make and in what sequence to arrange them. The proper way to read the above molecule, then, is in groups of three, as shown below:

> # Words to Know
>
> **Amino acid:** A relatively simple organic molecule from which proteins are made.
>
> **Deoxyribonucleic acid (DNA):** A large, complex molecule found in the nuclei of cells that carries genetic information.
>
> **Gene:** A section of a DNA molecule that carries instructions for the formation, functioning, and transmission of specific traits from one generation to another.
>
> **Mutagen:** Any substance or any form of energy that can bring about a mutation in DNA.
>
> **Nucleotide:** A unit from which DNA molecules are made.
>
> **Protein:** A complex chemical compound that consists of many amino acids attached to each other that are essential to the structure and functioning of all living cells.
>
> **Triad:** A group of three nucleotides in a DNA molecule that codes for the production of a single, specific amino acid.

-A-T-C - T-C-T - G-G-C - C-C-A - G-T-C - C-G-T - T-G-A - T-G-C-

But a DNA molecule can be damaged. A nucleotide might break loose from the DNA chain, a new nucleotide might be introduced into the chain, or one of the nucleotides in the chain might be changed. Suppose that the first of these possibilities occurred at the fifth nucleotide in the chain shown above. The result would be as follows:

-A-T-C - T- -T - G-G-C - C-C-A - G-T-C - C-G-T - T-G-A - T-G-C-

In this case, reading the nucleotides three at a time, as a cell always does, results in a different message than with the original chain. In the original chain, the nucleotide triads (sets of three nucleotides) are ATC TCT GGC CCA, and so on. But the nucleotide triads after the loss of one nucleotide are ATC TTG GCC CAG, and so on. The genetic message has changed. The cell is now instructed to make a different protein from the one it is supposed to make according to the original DNA code. A mutation has occurred.

Mutation

A mutation can also occur if a new nucleotide is introduced into the chain. Look at what happens when a new nucleotide, marked T*, is introduced into the original DNA chain:

-A-T-C - T-C-T - T*-G-G-C - C-C-A - G-T-C - C-G-T - T-G-A -

The nucleotide triads are now ATC TCT TGG CCC AGT, and so on. Again, a message different from the original DNA message is relayed.

Finally, a mutation can occur if a nucleotide undergoes a change. In the example below, the fifth nucleotide is changed from a C to a T:

-A-T-C - T-T-T - G-G-C - C-C-A - G-T-C - C-G-T - T-G-A - T-G-C-

It is obvious that the genetic message contained here is different from the original message.

Causes of mutation

Under most circumstances, DNA molecules are very stable. They survive in the nucleus of a cell without undergoing change, and they reproduce themselves during cell division without being damaged. But accidents do occur. For example, an X ray passing through a DNA molecule might break the chemical bond that holds two nucleotides together. The DNA molecule is destroyed and is no longer able to carry out its function.

A six-legged green frog. (Reproduced by permission of JLM Visuals.)

Anything that can bring about a mutation in DNA is called a mutagen. Most mutagens fall into one of two categories: They are either a form of energy or a chemical. In addition to X rays, other forms of radiation that can cause mutagens include ultraviolet radiation, gamma rays, and ionizing radiation. Chemical mutagens include aflatoxin (from mold), caffeine (found in coffee and colas), LSD (lysergic acid diethylamide; a hallucinogenic drug), benzo(a)pyrene (found in cigarette and coal smoke), Captan (a fungicide), nitrous oxide (laughing gas), and ozone (a major pollutant when in the lower atmosphere).

[*See also* **Carcinogen; Chromosome; Genetic disorders; Genetics; Human evolution**]

Natural gas

Natural gas is a fossil fuel. Most scientists believe natural gas was created by the same forces that formed oil, another fossil fuel. In prehistoric times, much of Earth was covered by water containing billions of tiny plants and animals that died and accumulated on ocean floors. Over the ages, sand and mud also drifted down to the ocean floor. As these layers piled up over millions of years, their weight created pressure and heat that changed the decaying organic material into oil and gas. In many places, solid rock formed above the oil and gas, trapping it in reservoirs.

Natural gas consists mainly of methane, the simplest hydrocarbon (organic compound that contains only carbon and hydrogen). It also contains small amounts of heavier, more complex hydrocarbons such as ethane, butane, and propane. Some natural gas includes impurities such as hydrogen sulfide ("sour" gas), carbon dioxide ("acid" gas), and water ("wet" gas). During processing, impurities are removed and valuable hydrocarbons are extracted. Sulfur and carbon dioxide are sometimes recovered and sold as by-products. Propane and butane are usually liquified under pressure and sold separately as LPG (liquified petroleum gas).

History of the discovery and use of natural gas

Natural gas is believed to have been first discovered and used by the Chinese, perhaps as early as 1000 B.C. Shallow stores of natural gas were released from just beneath the ground and piped short distances to be used as a fuel source. Natural gas provided a continuous source of

Natural gas

energy for flames. These "eternal fires" were found in temples and also used as attractions for visitors.

In 1821, an American gunsmith named William Aaron Hart drilled the first natural gas well in the United States. (To extract natural gas from the ground, a well must be drilled to penetrate the cap rock that covers it.) It was covered with a large barrel, and the gas was directed through wooden pipes that were replaced a few years later with lead pipe.

In the early 1900s, huge amounts of natural gas were found in Texas and Oklahoma, and in the 1920s modern seamless steel pipe was intro-

An offshore natural gas drilling platform. *(Reproduced by permission of The Stock Market.)*

> **Words to Know**
>
> **Fossil fuel:** Fuels formed by decaying plants and animals on the ocean floor that were covered by layers of sand and mud. Over millions of years, the layers of sediment created pressure and heat that helped bacteria change the decaying organic material into oil and gas.
>
> **Hydrocarbons:** Molecules composed solely of hydrogen and carbon atoms.

duced. The strength of this new pipe, which could be welded into long sections, allowed gas to be carried under higher pressures and, thus, in greater quantities. For the first time, natural gas transportation became profitable, and the American pipeline network grew tremendously through the 1930s and 1940s. By 1950, almost 300,000 miles (482,700 kilometers) of gas pipeline had been laid—a length greater than existing oil pipes.

Natural gas now supplies more than one-fourth of all energy consumed in America. In homes, natural gas is used in furnaces, stoves, water heaters, clothes dryers, and other appliances. The fuel also supplies energy for numerous industrial processes and provides raw materials for making many products that we use every day.

Natural gas and the environment

In light of environmental concerns, natural gas has begun to be reconsidered as a fuel for generating electricity. Natural gas is the cleanest burning fossil fuel, producing mostly just water vapor and carbon dioxide as by-products. Several gas power generation technologies have been advanced over the years, including a process that uses the principles of electrogasdynamics (EGD).

[*See also* **Gases, liquefaction of; Petroleum**]

Natural numbers

The natural numbers are the ordinary numbers, 1, 2, 3, etc., with which we count. They are sometimes called the counting numbers. They have

Natural numbers

been called natural because much of our experience from infancy deals with discrete (separate; individual; easily countable) objects such as fingers, balls, peanuts, etc. German mathematician Leopold Kronecker (1823–1891) is reported to have said, "God created the natural numbers; all the rest is the work of man."

Some disagreement exists as to whether zero should be considered a natural number. One normally does not start counting with zero. Yet zero does represent a counting concept: the absence of any objects in a set. To resolve this issue, some mathematicians define the natural numbers as the positive integers. An integer is a whole number, either positive or negative, or zero.

Operations involving natural numbers

Ultimately all arithmetic is based on the natural numbers. When multiplying 1.72 by .047, for example, the multiplication is done with the natural numbers 172 and 47. Then the result is converted to a decimal fraction by inserting a decimal point in the proper place. The placement of a decimal point is also done by counting natural numbers. When adding the fractions 1/3 and 2/7, the process is also one that involves natural numbers. First, the fractions are converted to 7/21 and 6/21. Then, the numerators are added using natural-number arithmetic, and the denominators copied. Even computers and calculators reduce their complex and lightning-fast computations to simple steps involving only natural numbers.

Measurements, too, are based on the natural numbers. In measuring an object with a meter stick, a person relies on the numbers printed near the centimeter marks to count the centimeters but has to physically count the millimeters (because they are not numbered). Whether the units are counted mechanically, electronically, or physically, the process is still one of counting, and counting is done with the natural numbers.

Number theory

One branch of mathematics concerns itself exclusively with the properties of natural numbers. This branch is known as number theory. Since the time of the ancient Greeks, mathematicians have explored these properties for their own sake and for their supposed connections with the supernatural. Most of this early research had little or no practical value. In recent times, however, many practical uses have been found for number theory. These include check-digit systems, secret codes, and other uses.

[*See also* **Arithmetic; Fraction, common; Number theory**]

Nautical archaeology

Nautical archaeology (pronounced NAW-tih-kul ar-kee-OL-low-jee) is the science of finding, collecting, preserving, and studying human objects that have become lost or buried under water. It is a fairly modern field of study since it depends primarily on having the technology both to locate submerged objects and to be able to remain underwater for some time to do real work. Whether it is conducted in freshwater or in the sea, and whether it finds sunken ships, submerged cities, or things deliberately thrown into the ocean, nautical archaeology is but another way of exploring and learning more about the human past.

Archaeology done underwater

Although some use the words nautical archaeology to mean a specialized branch of underwater archaeology, which is concerned only with ships and the history of seafaring, most consider the term to mean the same as the words underwater archaeology, undersea archaeology, marine archaeology, or maritime archaeology. All of these interchangeable terms mean simply that it is the study of archaeology being done underwater. Archaeology is the scientific study of the artifacts or the physical remains of past human cultures. By studying objects that ancient people have made, we can learn more about how they lived and even what they were like. In fact, studying ancient artifacts is the only way to learn anything about human societies that existed long before the invention of writing. For those later societies that are studied, being able to examine the actual objects made and used by those people not only adds to the written records they left behind, but allows us to get much closer to the reality of what life was like when they lived. Also, if we pay close attention to how the objects were made and used and what were their purposes, we begin to get a much more realistic picture of what these people were really like.

Underwater repositories of human history

Ever since the beginning of civilization and mankind's ability to move over water, the bottoms of nearly all oceans, lakes, and rivers became the final resting place for whatever those vessels were carrying. Once real trade began, it is safe to say that nearly every object made by humans was probably transported over water at some point in time, and just as frequent were mishaps and accidents of all sorts that resulted in those objects sinking to the bottom. Vessels of all types—from canoes,

Nautical archaeology

> **Words to Know**
>
> **Archaeology:** The scientific study of material remains, such as fossils and relics, of past societies.
>
> **Artifact:** In archaeology, any human-made item that relates to the culture under study.
>
> **Scuba:** A portable device including one or more tanks of compressed air used by divers to breathe underwater.

rafts, and barges to seafaring ships—became victims of every imaginable disaster. Vessels were sunk by severe weather and fierce storms, by construction defects and collisions, by robbery and warfare, by hidden sandbars and jagged reefs, and probably just as often by simple human error and misjudgment. Some cultures may have thrown things into the sea, perhaps to appease an angry god, while others conducted burials at sea. Finally, entire coastal cities are known to have been totally and permanently submerged as the result of an earthquake. All of these and more resulted in the creation of what might be called underwater repositories of human history.

Destroyed or preserved

Not all of these objects survived either the trip down to, or their stay on, the bottom. Their fate depended on where they landed. If an object sank near the seashore, chances are that it would have been broken by wave action. Even if it sank far below the action of waves, it still might not have survived, since it could have landed on submerged rocks and been broken by ocean currents. Sometimes underwater creatures, like snails and worms, burrowed inside and ate them, while others like coral or barnacles may have cemented themselves on the surface of an object and rotted or rusted away its inside.

However, besides hiding or destroying objects, the sea can also preserve them. Objects that sank into deep layers of mud were hidden from sight but were usually well-preserved. Often the saltiness of the water discouraged the growth of bacteria that can rot organic materials like wood. Other times, metals were buried in mud that allowed little or no air to get in, thus preventing them from corroding. It is not unusual, therefore, to discover ancient ships that have been deeply buried whose parts—from their wood boards to their ropes, masts, and nails—and cargos of pottery or weapons or even leather and cloth have been perfectly preserved.

Nautical archaeology

Underwater technology

People have been finding submerged objects of all sorts for as long as they have been able to get and stay below the surface. Early sponge divers were probably among the first, since they were expert at holding their breath and working underwater. Although primitive diving suits were used as early the sixteenth century, it was not until the nineteenth century that helmet diving gear was invented that allowed a person to "walk" on the bottom and explore it. Connected to the surface by an air hose and wearing what must have felt like a heavy suit of armor, the diver was clumsy and very slow and could never get very much done during his short trips to the bottom.

Nautical fossils are examined in much the same way as fossils found on dry land. (Reproduced by permission of The Corbis Corporation [Bellevue].)

Nautical archaeology

Nautical archaeology did not become a feasible pursuit until the invention in 1943 of an underwater breathing device by French naval officer and ocean explorer Jacques-Yves Cousteau (1910–1997) and Emile Gagnan, also of France. Called scuba gear for self-contained underwater breathing apparatus (and trademarked under the name Aqua-Lung), it revolutionized diving and allowed a person to swim freely down to about 180 feet (55 meters) wearing only a container of highly compressed air on his back. It was later improved by using a mixture of oxygen and helium rather than normal air (which is oxygen and nitrogen), and this allowed a diver to descend as deep as 1,640 feet (500 meters). Until this invention, actual underwater exploring had been done mostly by professional divers who were directed by archaeologists. With this new scuba gear, however, archaeologists could explore themselves. From this, modern nautical archaeology was born.

Improving technology

The first underwater site to be excavated (exposed by digging) by diving archaeologists was a Bronze Age (c. 1200 B.C.) ship wrecked off the coast of Turkey. It was explored by Americans Peter Throckmorton and George Bass, who became pioneers in the field. They and all others to follow used nearly the same techniques that archaeologists on land al-

This fossilized spadefish is over 50 million years old. *(Reproduced by permission of The Corbis Corporation [Bellevue].)*

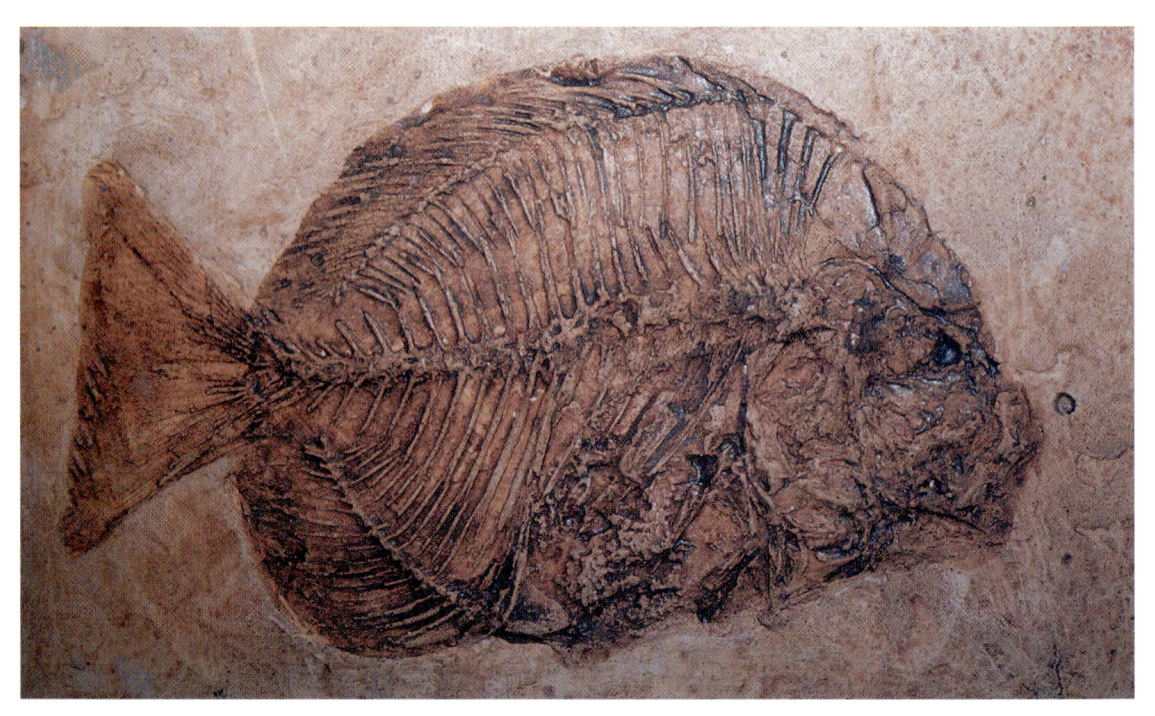

ways follow, although working underwater made their job one of the most difficult and demanding of all scientific activities.

Today, nautical archaeologists employ a variety of technologies and techniques that make their job easier. They sometimes use aerial photographs to get detailed pictures of shallow, clear water. They often use metal detectors or a magnetometer (pronounced mag-neh-TAH-meh-ter) to find metal objects. Sonar devices send waves of sound through the water that bounce off solid objects and return as echoes, which are recorded by electronic equipment. Underwater cameras are regularly used, as are remotely operated vehicles that can penetrate to extreme depths where severe cold, high pressure, and total darkness would prevent humans from going. Finally, before excavating, nautical archaeologists carefully study and map a site (the location of a deposit or a wreck). This is probably the most time-consuming part of the job, as each artifact is drawn on a map to note its exact location. Only after the entire site is mapped will removal begin. This is done using several different methods. Balloons or air bags are often used to raise large or heavy objects. Vacuum tubes called airlifts are used to suck up smaller objects or pieces. Certain objects brought to the surface must be properly cared for or they can fall apart in a matter of days. Nautical archaeologists must therefore have ready a thorough plan to preserve these fragile objects once they are raised.

Nautical archaeology is still a young science, but it has achieved some spectacular results. Entire ships, like the Swedish warship *Vasa*, which sank in 1628, and the even older English ship *Mary Rose*, have been raised. The *Vasa* took five years to raise; the *Mary Rose* took nearly twice that long. The wreck of the *Titanic*, which sunk in 1912 after hitting an iceberg, has been thoroughly explored ever since it was first located by a remote-control submarine in 1985. As technology improves, so does the ability of nautical archaeologists to explore the hidden museum under the sea that holds more clues about our human past.

[*See also* **Archaeology**]

Nebula

Bright or dark clouds hovering in the interstellar medium (the space between the stars) are called nebulae. *Nebula,* Latin for "cloud," is a visual classification rather than a scientific one. Objects called nebulae vary greatly in composition. Some are really galaxies, but to early astronomers they all appeared to be clouds.

Nebula

> ### Words to Know
>
> **Cepheid variable:** Pulsating yellow supergiant star that can be used to measure distance in space.
>
> **Infrared radiation:** Electromagnetic radiation of a wavelength shorter than radio waves but longer than visible light that takes the form of heat.
>
> **Interstellar medium:** Space between the stars, consisting mainly of empty space with a very small concentration of gas atoms and tiny solid particles.
>
> **Light-year:** Distance light travels in one solar year, roughly 5.9 trillion miles (9.5 trillion kilometers).
>
> **Red giant:** Stage in which an average-sized star (like our sun) spends the final 10 percent of its lifetime; its surface temperature drops and its diameter expands to 10 to 1,000 times that of the Sun.
>
> **Stellar nursery:** Area within glowing clouds of dust and gas where new stars are being formed.
>
> **Supernova:** Explosion of a massive star at the end of its lifetime, causing it to shine more brightly than the rest of the stars in the galaxy put together.
>
> **Ultraviolet radiation:** Electromagnetic radiation of a wavelength just shorter than the violet (shortest wavelength) end of the visible light spectrum.

Bright nebulae

Some categories of bright nebulae include spiral, planetary, emission, and reflection. Others are remnants of supernova explosions.

In 1923, American astronomer Edwin Hubble made a remarkable discovery about a spiral-shaped nebula: it was actually a gigantic spiral galaxy. Previously, astronomers had considered the Great Nebula in the constellation Andromeda to be a cloud of gas within our galaxy, the Milky Way. Hubble identified a variable star known as a Cepheid (pronounced SEF-ee-id; a blinking star used to measure distance in space) in the Andromeda nebula, estimating its distance to be about one million light-years away. This was far beyond the bounds of the Milky Way, proving the existence of galaxies outside of our own. Since then, many other spiral nebulae have been defined as galaxies.

Planetary nebulae truly are clouds of gas. They are called planetary because when viewed through a telescope, they appear greenish and round, like planets. Astronomers believe a planetary nebula is a star's detached outer atmosphere of hydrogen gas. This is a by-product of a star going through the later stages of its life cycle. As it evolves past the red giant stage, a star sheds its atmosphere, much like a snake sheds its skin. One of the most famous of these is the Ring Nebula in the constellation Lyra.

An emission nebula is a glowing gas cloud with a hot bright star within or behind it. The star gives off high-energy ultraviolet radiation, which ionizes (electrically charges) the gas. As the electrons recombine with the atoms of gas, the gas fluoresces, or gives off light. A well-known example is the Orion Nebula, a greenish, hydrogen-rich, star-filled cloud

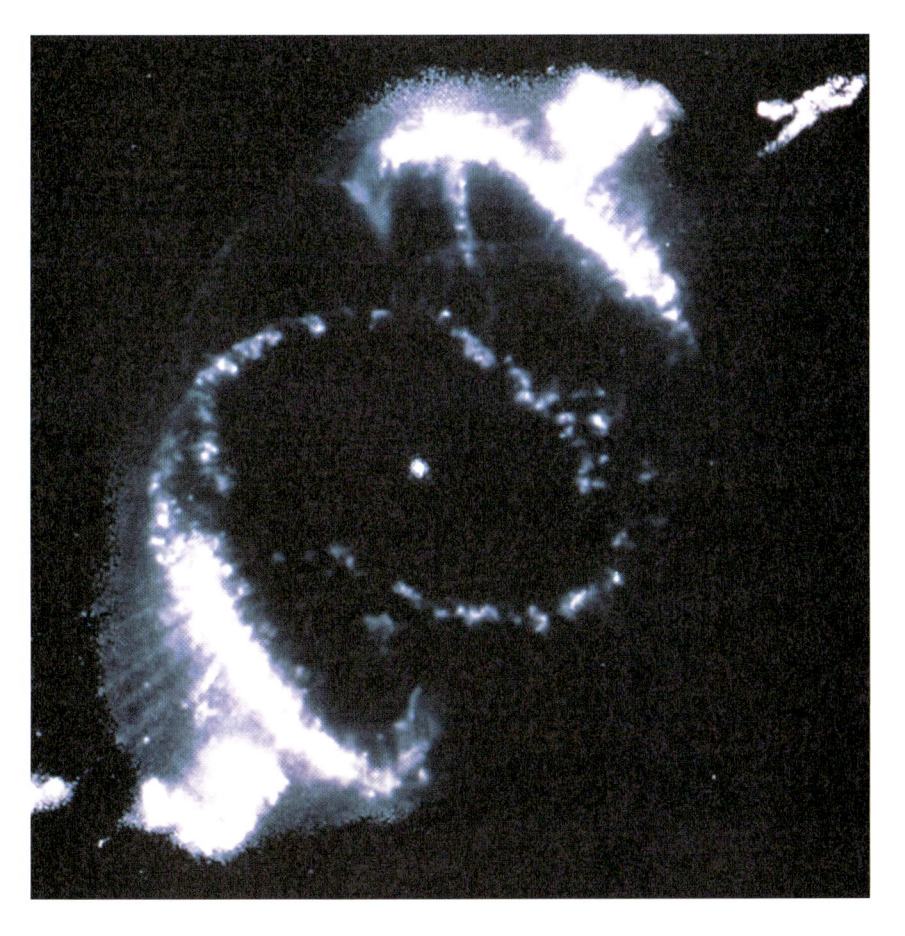

The Cat's Eye Nebula as seen from the Hubble Space Telescope. At center is a dying star during its last stages of life. Knots and thin filaments can be seen along the edge of the gas. *(Reproduced by permission of National Aeronautics and Space Administration.)*

that is 20 light-years across. Astronomers believe it to be a stellar nursery, a place where new stars are formed.

Reflection nebulae are also bright gas clouds, but not as common as emission nebulae. A reflection nebula is a bluish cloud containing dust that reflects the light of a neighboring bright star. It is blue for a similar reason that Earth's sky is blue. In the case of our sky, the blue wavelength of sunlight is scattered by gas molecules in our atmosphere. In the same way, the nebula's dust scatters starlight only in the wavelengths of blue light.

The final type of bright nebula is that produced by a supernova explosion. The most famous nebula of this type is the Crab Nebula, an enormous patch of light in the constellation Taurus. At its center lies a pulsar, a rapidly spinning, incredibly dense star made of neutrons that remains after a supernova explosion.

Dark nebulae

Dark nebulae are also scattered throughout the interstellar medium. They appear dark because they contain dust (composed of carbon, silicon, magnesium, aluminum, and other elements) that does not emit light and that is dense enough to block the light of stars beyond. These nonglowing clouds are not visible through an optical telescope, but do give off infrared radiation. They can thus be identified either as dark patches on a background of starlight or through an infrared telescope. One example of a dark nebula is the cloud that blots out part of the Cygnus constellation in our galaxy.

[*See also* **Infrared astronomy; Interstellar matter**]

Neptune

Neptune, the eighth planet away from the Sun, was discovered in 1846 by German astronomer Johann Galle, who based his finding on the mathematical predictions of French astronomer Urbain Le Verrier and English astronomer John Couch Adams. Because Neptune is so far way from the Sun—about 2.8 billion miles (4.5 billion kilometers)—it is difficult to observe. Very little was known about it until fairly recently. In August 1989, the U.S. space probe *Voyager 2* flew by Neptune, finally providing some answers about this mysterious, beautiful globe.

Neptune is a large planet, with a mass 17 times that of Earth. The diameter at its equator is roughly 30,700 miles (49,400 kilometers). Neptune spins slightly faster than Earth—its day is equal to just over 19 Earth hours. It completes one revolution around the Sun in about 165 Earth years.

Since it is the color of water, Neptune was named for the Roman god of the sea. Its blue-green color, however, is due to methane gas. The thick outer atmospheric layer of hydrogen, helium, and methane is extremely cold: −350°F (−212°C). Below the atmosphere lies an ocean of ionized (electrically charged) water, ammonia, and methane ice. Underneath the ocean, which reaches thousands of miles in depth, is a rocky iron core.

Neptune is seventeen times larger than Earth. *(Reproduced by permission of National Aeronautics and Space Administration.)*

Neptune

Neptune is subject to the fiercest winds in the solar system. It has a layer of blue surface clouds that whip around with the wind and an upper layer of wispy white clouds of methane crystals that rotate with the planet. At the time of *Voyager 2*'s encounter, three storm systems were evident on its surface. The most prominent was a dark blue area called the Great Dark Spot, which was about the size of Earth. Another storm, about the size of our moon, was called the Small Dark Spot. Then there was Scooter, a small, fast-moving white storm system that seemed to chase the other storms around the planet. Its true nature remains a mystery.

In 1994, however, observations from the Hubble Space Telescope showed that the Great Dark Spot had disappeared. Astronomers theorize the spot either simply dissipated or is being masked by other aspects of the atmosphere. A few months later, the Hubble Space Telescope discovered a new dark spot in Neptune's northern hemisphere. This discovery has led astronomers to conclude that the planet's atmosphere changes rapidly, which might be due to slight changes in the temperature differences between the tops and bottoms of the clouds.

Neptune's magnetic field

A magnetic field has been measured on Neptune, tilted from its axis at a 48-degree angle and just missing the center of the planet by thousands of miles. This field is created by water beneath the surface that measures 4,000°F (2,204°C), water so hot and under so much pressure that it generates an electrical field.

Voyager 2 found that Neptune is encircled by at least four very faint rings, much less pronounced than the rings of Saturn, Jupiter, or Uranus. Although astronomers are not quite sure, they believe these rings are composed of particles, some of which measure over a mile across and are considered moonlets. These particles clump together in places, creating relatively bright arcs. This originally led astronomers to believe that only arcs—and not complete rings—were all that surrounded the planet.

The moons of Neptune

Neptune has eight moons, six of which were discovered by *Voyager 2*. The largest, Triton, was named for the son of the mythical Neptune. Triton was discovered a month after Neptune itself. It is 1,681 miles (3,705 kilometers) in diameter and has a surface temperature of −400°F (−240°C), making it the coldest place in the solar system. It has a number of unusual qualities. First, this peach-colored moon orbits Neptune in the opposite direction of all the other planets' satellites, and it rotates on its axis in the opposite direction that Neptune rotates. In addition, *Voyager* found that Triton has an atmosphere with layers of haze, clouds, and wind streaks. All of this information has led astronomers to conclude that Triton was captured by Neptune long ago from an independent orbit around the Sun.

The second Neptunian moon, a faint, small body called Nereid, was discovered in 1949 by Dutch astronomer Gerald Kuiper. The other six moons range from 30 miles (50 kilometers) to 250 miles (400 kilometers) in diameter.

[*See also* **Solar system; Space probe**]

Nervous system

The nervous system is a collection of cells, tissues, and organs through which an organism receives information from its surroundings and then directs the organism as to how to respond to that information. As an example, imagine that a child accidentally touches a very hot piece of metal. The cells in the child's hand that detect heat send a message to the child's brain. The brain receives and analyzes that message and sends back a message to the child's hand. The message tells the muscles of the hand to pull itself away from the heat.

The basic unit of the nervous system is a neuron. A neuron is a nerve cell capable of passing messages from one end to the other. In the example above, the "hot" message was passed from one neuron to the next along a path that runs from the child's hand to its brain. The "move your hand" message then passed from one neuron to the next along another path running from the child's brain back to its hand.

Types of nervous systems

The complexity of nervous systems differs from organism to organism. In the simplest of organisms, the nervous system may consist of little more than a random collection of neurons. Such systems are known as a nerve net. An example of an animal with a nerve net is the hydra, a cylinder-shaped freshwater polyp. Hydra respond to stimuli such as heat, light, and touch, but their nerve net is not a very effective way to transmit messages. Their responses tend to be weak and localized.

In other organisms, neurons are bunched together in structures known as ganglia (single: ganglion). Flatworms, for example, have a pair of ganglia that function like a simple brain. The ganglia are attached to

Nervous system

> **Words to Know**
>
> **Autonomic nervous system:** A collection of neurons that carry messages from the central nervous system to the heart, smooth muscles, and glands generally not as a result of conscious action on the part of the brain.
>
> **Central nervous system:** The portion of the nervous system in a higher organism that consists of the brain and spinal cord.
>
> **Ganglion:** A bundle of neurons that acts something like a primitive brain.
>
> **Motor neutrons:** Neurons that carry messages from the central nervous system to muscle cells.
>
> **Nerve net:** A simple type of nervous system consisting of a random collection of neurons.
>
> **Neuron:** A nerve cell.
>
> **Parasympathetic nervous system:** A collection of neurons that control a variety of internal functions of the body under normal conditions.
>
> **Peripheral nervous system:** The portion of the nervous system in an organism that consists of all the neurons outside the central nervous system.
>
> **Sensory neurons:** Neurons that respond to stimuli from an organism's surroundings.
>
> **Somatic nervous system:** A collection of neurons that carries messages from the central nervous system to muscle cells.
>
> **Stimuli:** Something that causes a response.
>
> **Sympathetic nervous system:** A collection of neurons that control a variety of internal functions when the body is exposed to stressful conditions.

two nerve cords that run the length of the worm's body. These two cords are attached to each other by other nerves. This kind of nervous system is sometimes described as a ladder-type nervous system.

The human nervous system. The most complex nervous systems are found in the vertebrates (animals with backbones), including humans. These nervous systems consist of two major divisions: the central nervous system and the peripheral nervous system. The central nervous system consists of the brain and spinal cord, and the peripheral system of all neurons outside the central nervous system. The brains of different vertebrate species differ from each other in their size and complexity, but all contain three general areas, known as the forebrain, midbrain, and hindbrain. These areas look different, however, and have somewhat different functions in various species.

The peripheral nervous system consists of two kinds of neurons known as sensory neurons and motor neurons. Sensory neurons are located in the sensory organs, such as the eye and ear. They are able to detect stimuli from outside the organism, such as light or sound. They then pass that information through the peripheral nervous system to the spinal cord and then on to the brain. Motor neurons carry messages from the brain, through the spinal cord, and to the muscles. They tell certain muscles to contract in order to respond to stimuli in some way or another.

The peripheral nervous system can be subdivided into two parts: the somatic system and the autonomic system. The somatic system involves the skeletal muscles. It is considered to be a voluntary system since the brain exerts control over movements such as writing or throwing a ball. The autonomic nervous system affects internal organs, such as the heart, lungs, stomach, and liver. It is considered to be an involuntary system since the processes it controls occur without conscious effort on the part

A scanning electron micrograph of three neurons in the human brain. *(Reproduced by permission of Photo Researchers, Inc.)*

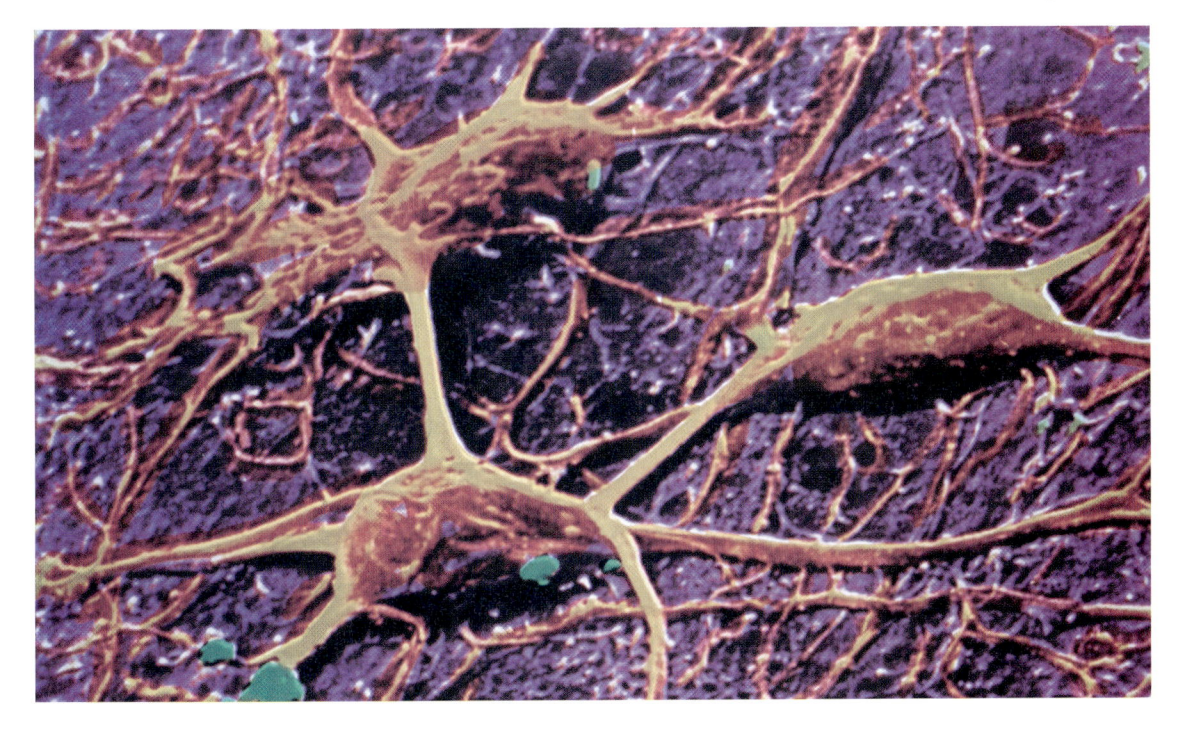

Nervous system

> ## Pain
>
> Where would humans be without pain? We feel pain when we put a finger into a flame or touch a sharp object. What would happen if our body did not recognize what had happened? What would happen if we left our finger in the flame or did not pull away from the sharp object? Pain is obviously a way that organisms have evolved for protecting themselves from dangerous situations.
>
> Although the reality of pain is well known to everyone, scientists still know relatively little as to how pain actually occurs. Current theories suggest that a "painful" event results in the release of certain "pain message" chemicals. These chemicals travel through the peripheral nervous system and into the central nervous system. Within the spinal cord and the brain, those pain messages are analyzed and an appropriate response is prepared. For example, the arrival of a pain message in the spinal cord is thought to result in the release of chemicals known as endorphins and enkephalins. These compounds are then thought to travel back to the sensory neurons and prevent the release of any additional pain message chemicals.
>
> An interesting feature of pain is the role that individual psychology plays. For example, some people seem to be more sensitive to pain than others. This difference may reflect factors such as fear, expectation, upbringing, and emotions as much as physical factors. Another interesting phenomenon is phantom pain. Phantom pain is the pain a person feels in an amputated limb. How can pain continue when the site of that pain is no longer there? These and many other questions remain in the search to discover those bodily changes that occur during pain, the reasons that pain occurs, and the ways in which pain can be eliminated.

of an individual. For example, we do not need to think about digesting our food in order for that event to take place.

The autonomic nervous system is itself divided into two parts: the parasympathetic and sympathetic systems. The parasympathetic system is active primarily in normal, restful situations. It acts to decrease heartbeat and to stimulate the movement of food and the secretions necessary for digestion. The sympathetic nervous system is most active during times of stress and becomes dominant when the body needs energy. It increases the rate and strength of heart contractions and slows down the process of digestion. The sympathetic and parasympathetic nervous systems are said to operate antagonistically. In other words, when one system is dominant, the other is quiet.

Neuromuscular diseases

Nerves and muscles usually work together so smoothly that we don't even realize what is happening. Messages from the brain carry instructions to motor neurons, telling them to move in one way or another. Whenever we walk, talk, smile, turn our head, or pick up a pencil, our nervous and muscular systems are working in perfect harmony.

But this smooth combination can break down. Nerve messages do not reach motor neurons properly, or those neurons do not respond as they have been told to respond. The result of such break downs is a neuromuscular disease. Perhaps the best known example of such disorders is muscular dystrophy (MD). The term muscular dystrophy actually applies to a variety of closely related conditions. The most common form of muscular dystrophy is progressive (or Duchenne) muscular dystrophy.

Progressive muscular dystrophy is an inherited disorder that affects males about five times as often as females. It occurs in approximately 1 out of every 3,600 newborn males. The condition is characterized by weakness in the pelvis, shoulders, and spine and is usually observed by the age of five. The condition becomes more serious with age, and those who inherit MD seldom live to maturity.

The causes of other forms of muscular dystrophy and other neuromuscular disorders are not well known. They continue to be, however, the subject of intense research by medical scientists.

[*See also* **Brain; Muscular system; Neuron**]

Neutron

A neutron is one of two particles found inside the nucleus (central part) of an atom. The other particle is called a proton. Electrons are particles that move around an atom outside the nucleus.

Discovery of the atom

British physicist Ernest Rutherford discovered the atom in 1911. He constructed a model showing an atom with a nucleus containing protons

Neutron

> **Words to Know**
>
> **Axon:** The projection of a neuron that carries an impulse away from the cell body of the neuron.
>
> **Central nervous system:** The portion of the nervous system in a higher organism that consists of the brain and spinal cord.
>
> **Cytoplasm:** The fluid inside a cell that surrounds the nucleus and other membrane-enclosed compartments.
>
> **Dendrite:** A portion of a nerve cell that carries nerve impulses toward the cell body.
>
> **Ion:** A molecule or atom that has lost one or more electrons and is, therefore, electrically charged.
>
> **Myelin sheath:** A white, fatty covering on nerve axons.
>
> **Neurotransmitter:** A chemical used to send information between nerve cells or nerve and muscle cells.
>
> **Peripheral nervous system:** The portion of the nervous system in an organism that consists of all the neurons outside the central nervous system.
>
> **Receptors:** Locations on cell surfaces that act as signal receivers and allow communication between cells.
>
> **Stimulus:** Something that causes a response.
>
> **Synapse:** The space between two neurons through which neurotransmitters travel.

and electrons. Scientists studying the model knew that something must be missing from it. Rutherford suggested that some sort of neutral particle might exist in the nucleus. He and a graduate student working with him, James Chadwick, could not prove his theory, mainly because neutrons cannot be detected by any standard tools such as cloud chambers or Geiger counters.

Finally, Chadwick tried directing a beam of radiation at a piece of paraffin (a waxy mixture used to make candles). He observed that protons were ejected from the paraffin. Chadwick concluded that the radiation must consist of particles with no charge and a mass about equal to that of the proton. That particle was the neutron.

In the early 1960s, the American physicist Robert Hofstadter discovered that both protons and neutrons contain a central core of positively charged matter that is surrounded by two shells. In the neutron, one shell is negatively charged, just balancing the positive charge in the particle's core.

[*See also* **Alzheimer's disease; Nervous system**]

Neutron star

A neutron star is the dead remnant of a massive star. A star reaches the end of its life when it uses up all of its nuclear fuel. Without fuel, it cannot undergo nuclear fusion, the process that pushes matter outward from the star's core and provides a balance to its immense gravitational field. The fate of a dying star, however, depends on that star's mass.

A medium-sized star, like the Sun, will shrink and end up as a white dwarf (small, extremely dense star having low brightness). The largest stars—those more than three times the mass of the Sun—explode in a supernova and then, in theory, undergo a gravitational collapse so complete they form black holes (single points of infinite mass and gravity). Those stars larger than the Sun yet not more than three times its mass will also explode in a supernova, but will then cave in on themselves to form a densely packed neutron star.

Origin of a neutron star

A neutron star is formed in two stages. First, within a second after nuclear fusion on the star's surface ceases, gravity crushes the star's atoms. This forces protons (positively charged particles) and electrons (negatively charged particles) together to form neutrons (uncharged particles) and expels high-energy subatomic particles called neutrinos. The star's core, which started out about the size of Earth, is compacted into a sphere less than 60 miles (97 kilometers) across.

In the second stage, the star undergoes a gravitational collapse and then, becoming energized by the neutrino burst, explodes in a brilliant supernova. All that remains is an extremely dense neutron core, about 12 miles (19 kilometers) in diameter with a mass nearly equal to that of the Sun. A sugar-cube-sized piece of neutron star would weigh billions of tons.

Neutron stars spin rapidly. This is because the original stellar core was spinning as it collapsed, naturally increasing its rate of spin.

Neutron star

> **Words to Know**
>
> **Black hole:** Remains of a massive star after it has exploded in a supernova and collapsed under tremendous gravitational force into a single point of infinite mass and gravity.
>
> **Neutrino:** A subatomic particle resulting from certain nuclear reactions that has no charge and possibly no mass.
>
> **Nuclear fusion:** Process in which the nuclei of two hydrogen atoms are fused together at extremely high temperatures to form a single helium nucleus, releasing large amounts of energy as a by-product.
>
> **Pulsar:** Rapidly rotating neutron star that emits varying radio waves at precise intervals.
>
> **Radiation:** Energy transmitted in the form of electromagnetic waves or subatomic particles.
>
> **Subatomic particle:** Basic unit of matter and energy (proton, neutron, electron, neutrino, and positron) smaller than an atom.
>
> **Supernova:** Explosion of a massive star at the end of its lifetime, causing it to shine more brightly than the rest of the stars in the galaxy put together.
>
> **White dwarf:** Small, extremely dense star having low brightness.

Neutron stars also have intense gravitational and magnetic fields. The gravity is strong because there is so much matter packed into so small an area. The spinning generates a magnetic field, and the star spews radiation out of its poles like a lighthouse beacon. Neutron stars give off radiation in a variety of wavelengths: radio waves, visible light, X rays, and gamma rays.

Pulsars

If the magnetic axis of the neutron star is tilted a certain way, the spinning star's on-and-off signal can be detected from Earth. This fact led to the discovery of the first neutron star in 1967 by English astronomer Antony Hewish and his student Jocelyn Bell Burnell.

Hewish and Bell Burnell were conducting an experiment to track quasars (extremely bright, distant objects) when they discovered a mysterious, extremely regular, pulsing signal. They found similar signals coming from other parts of the sky, including one where a supernova was known to have occurred. With the help of astronomer Thomas Gold, they learned that the signals matched the predicted pattern of neutron stars. They named these blinking neutron stars pulsars (from pulsating stars).

Since then, more than 500 pulsars have been catalogued, including many in spots where a supernova is known to have occurred. Pulse rates of observed neutron stars range from 4 seconds to 1.5 milliseconds. Scientists believe that more than 100,000 active pulsars may exist in our galaxy.

[*See also* **Star; Subatomic particles; Supernova**]

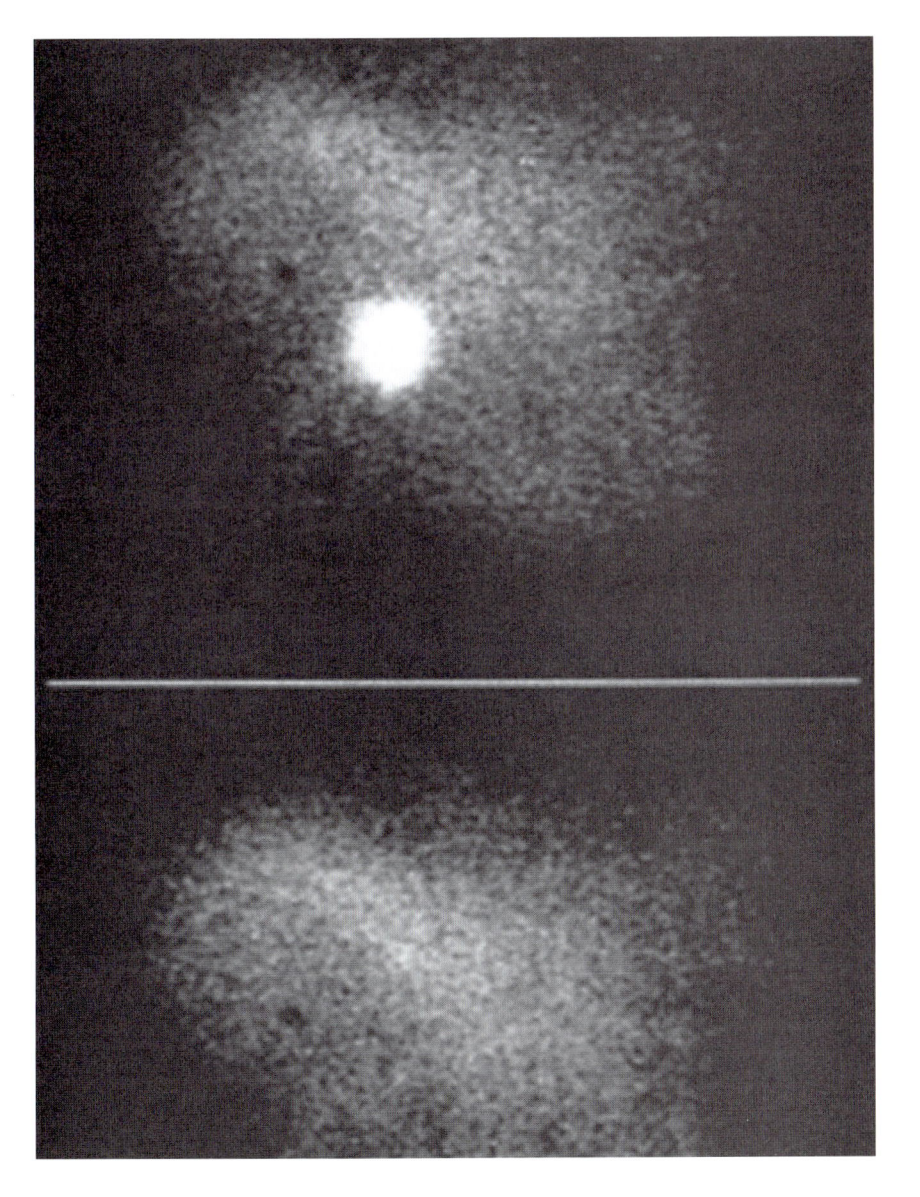

X-ray images showing the neutron star at the heart of the Crab Nebula. The remnant of a supernova seen from Earth in 1954, this neutron star emits radiation in bursts—appearing to blink on and off—and thus is a pulsar. *(Reproduced by permission of National Aeronautics and Space Administration.)*

Nitrogen cycle

The term nitrogen cycle refers to a series of reactions in which the element nitrogen and its compounds pass continuously through Earth's atmosphere, lithosphere (crust), and hydrosphere (water component). The major components of the nitrogen cycle are shown in the accompanying figure. In this diagram, elemental nitrogen is represented by the formula N_2, indicating that each molecule of nitrogen consists of two nitrogen atoms. In this form, nitrogen is more correctly called dinitrogen.

Nitrogen fixation

Nitrogen is the most abundant single gas in Earth's atmosphere. It makes up about 80 percent of the atmosphere. This fact is important because plants require nitrogen for their growth and, in turn, animals depend on plants for their survival. The problem is, however, that plants are unable to use nitrogen in its elemental form—as dinitrogen. Any process by which elemental dinitrogen is converted to a compound is known as nitrogen fixation.

The nitrogen cycle. *(Reproduced by permission of The Gale Group.)*

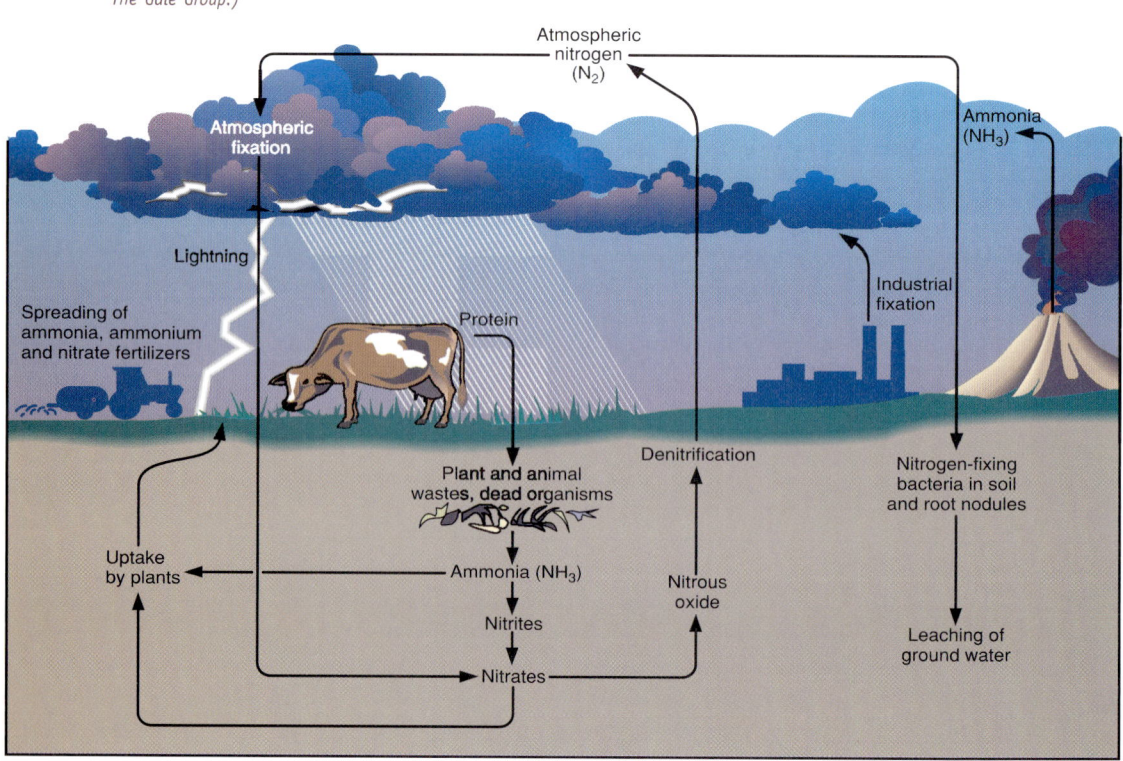

Nitrogen cycle

Words to Know

Ammonification: The conversion of nitrogen compounds from plants and animals to ammonia and ammonium; this conversion occurs in soil or water and is carried out by bacteria.

Denitrification: The conversion of nitrates to dinitrogen (or nitrous oxide) by bacteria.

Dinitrogen fixation (nitrogen fixation): The conversion of elemental dinitrogen (N2) in the atmosphere to a compound of nitrogen deposited on Earth's surface.

Nitrification: The process by which bacteria oxidize ammonia and ammonium compounds to nitrites and nitrates.

Dinitrogen is converted from an element to a compound by a number of naturally occurring processes. When lightning passes through the atmosphere, it prompts a reaction between nitrogen and oxygen; oxides of nitrogen—primarily nitric oxide (NO) and nitrogen dioxide (NO_2)—are formed. Both oxides then combine with water vapor in the atmosphere to form nitric acid (HNO_3). Nitric acid is carried to the ground in rain and snow, where it is converted to nitrites and nitrates. Nitrites and nitrates are both compounds of nitrogen and oxygen, the latter containing more oxygen than the former. Naturally occurring minerals such as saltpeter (potassium nitrate; KNO_3) and Chile saltpeter (sodium nitrate; $NaNO_3$) are the most common nitrates found in Earth's crust.

Certain types of bacteria also have the ability to convert elemental dinitrogen to nitrates. Probably the best known of these bacteria are the *rhizobium,* which live in nodules on the roots of leguminous plants such as peas, beans, clover, and the soya plant.

Finally, dinitrogen is now converted to nitrates on very large scales by human processes. In the Haber process, for example, nitrogen and hydrogen are combined to form ammonia, which is then used in the manufacture of synthetic fertilizers, most of which contain nitrates.

Ammonification, nitrification, and denitrification

Nitrogen that has been fixed by one of the mechanisms described above can then be taken in by plants through their roots and used to build new stems, leaves, flowers, and other structures. Almost all animals obtain the nitrogen they require, in turn, by eating plants and taking in the plant's organic forms of nitrogen.

The nitrogen stored in plants and animals is eventually returned to Earth by one of two processes: elimination (in the case of animals) or death (in the case of both animals and plants). In whatever form the nitrogen occurs in the dead plant or animal, it is eventually converted to ammonia (NH_3) or one of its compounds. Compounds formed from ammonia are known as ammonium compounds. This process of ammonification is carried out (as the plant or animal decays) by a number of different microorganisms that occur naturally in the soil.

Ammonia and ammonium compounds, in their turn, are then converted to yet another form, first to nitrites and then to nitrates. The transformation of ammonia and ammonium to nitrite and nitrate is an oxidation process that takes place through the action of various bacteria such as those in the genus *Nitrosomonas* and *Nitrobacter*. The conversion of ammonia and ammonium compounds to nitrites and nitrates is called nitrification.

In the final stage of the nitrogen cycle, oxygen is removed from nitrates by bacteria in a process known as denitrification. Denitrification converts nitrogen from its compound form to its original elemental form as dinitrogen, and the cycle is ready to begin once again.

Nitrogen family

The nitrogen family consists of the five elements that make up Group 15 of the periodic table: nitrogen, phosphorus, arsenic, antimony, and bismuth. These five elements share one important structural property: they all have five electrons in the outermost energy level of their atoms. Nonetheless, they are strikingly different from each other in both physical properties and chemical behavior. Nitrogen is a nonmetallic gas; phosphorus is a solid nonmetal; arsenic and antimony are metalloids; and bismuth is a typical metal.

Nitrogen

Nitrogen is a colorless, odorless, tasteless gas with a melting point of −210°C (−346°F) and a boiling point of −196°C (−320°F). It is the most abundant element in the atmosphere, making up about 78 percent by volume of the air that surrounds Earth. The element is much less common in Earth's crust, however, where it ranks thirty-third (along with gallium) in abundance. Scientists estimate that the average concentration of nitrogen in crustal rocks is about 19 parts per million, less than that of elements such as neodymium, lanthanum, yttrium, and scandium, but greater than that of well-known metals such as lithium, uranium, tungsten, silver, mercury, and platinum.

The most important naturally occurring compounds of nitrogen are potassium nitrate (saltpeter), found primarily in India, and sodium nitrate (Chile saltpeter), found primarily in the desert regions of Chile and other parts of South America. Nitrogen is also an essential component of the proteins found in all living organisms.

Credit for the discovery of nitrogen in 1772 is usually given to Scottish physician Daniel Rutherford (1749–1819). Three other scientists, Henry Cavendish, Joseph Priestley, and Carl Scheele, could also claim to have discovered the element at about the same time. Nitrogen was first identified as the product left behind when a substance is burned in a closed sample of air (which removed the oxygen component of air).

Uses. The industrial uses of nitrogen have increased dramatically in the past few decades. It now ranks as the second most widely produced chem-

Computer graphics representation of a diatomic molecule of nitrogen. (Diatomic means there are two atoms in the molecule.) The spheres off center are the nitrogen atoms, and the area between the atoms represents their strong bond. *(Reproduced by permission of Photo Researchers, Inc.)*

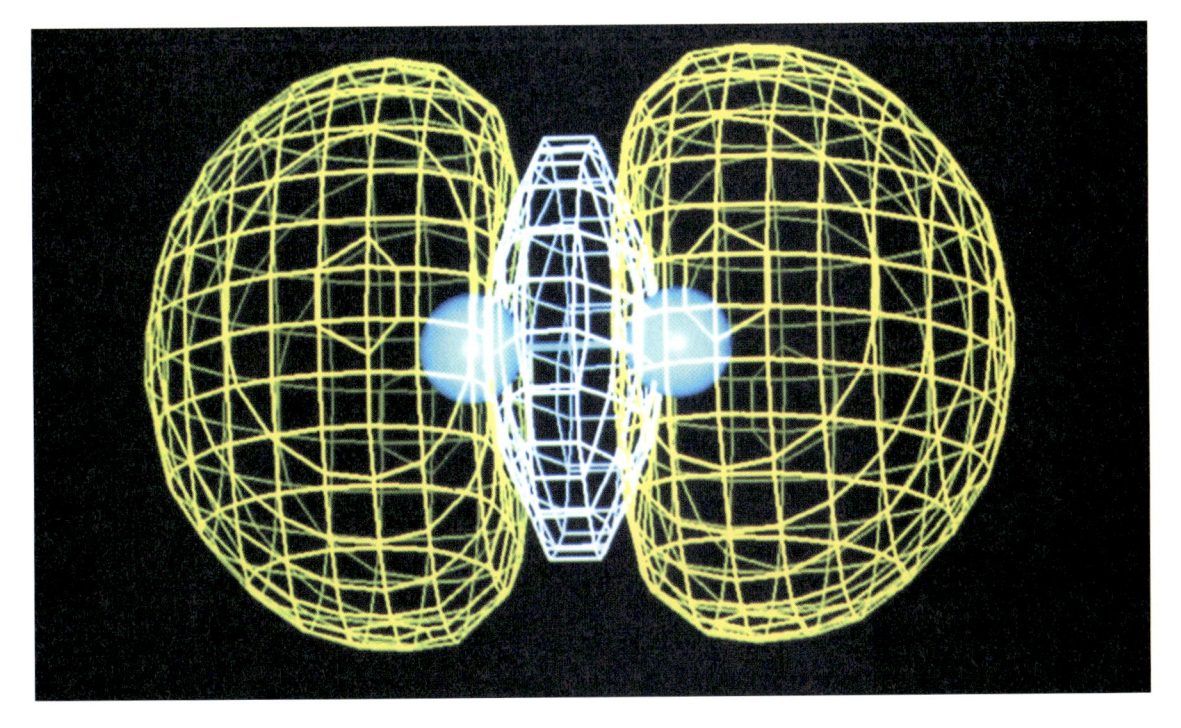

ical in the United States with an annual production of about 57 billion pounds (26 billion kilograms).

The element's most important applications depend on its chemical inertness (inactivity). It is widely used as a blanketing atmosphere in metallurgical processes where the presence of oxygen would be harmful. In the processing of iron and steel, for example, a blanket of nitrogen placed above the metals prevents their reacting with oxygen, which would form undesirable oxides in the final products.

The purging (freeing of sediment or trapped air) of tanks, pipes, and other kinds of containers with nitrogen can also prevent the possibility of fires. In the petroleum industry, for example, the processing of organic compounds in the presence of air creates the potential for fires—fires that can be avoided by covering the reactants with pure nitrogen.

Nitrogen is also used in the production of electronic components. Assembly of computer chips and other electronic devices can take place with all materials submerged in a nitrogen atmosphere, preventing oxidation of any of the materials in use. Nitrogen is often used as a protective agent during the processing of foods so that decay (oxidation) does not occur.

Another critical use of nitrogen is in the production of ammonia by the Haber process, named after its inventor, German chemist Fritz Haber (1868–1934). The Haber process involves the direct synthesis of ammonia from its elements—nitrogen and hydrogen. The two gases are combined under specific conditions: (1) the temperature must be 500 to 700°C (900 to 1300°F), (2) the pressure must be several hundred atmospheres, and (3) a catalyst (something that speeds up chemical reactions) such as finely divided nickel must be present. One of the major uses of the ammonia produced by this method is in the production of synthetic fertilizers.

About one-third of all nitrogen produced is used in its liquid form. For example, liquid nitrogen is used for quick-freezing foods and for preserving foods in transit. Additionally, the very low temperatures of liquid nitrogen make some materials easier to handle. For example, most forms of rubber are too soft and pliable for machining at room temperature. They can, however, first be cooled in liquid nitrogen and then handled in a much more rigid form.

Three compounds of nitrogen are also commercially important and traditionally rank among the top 25 chemicals produced in the United States. They are ammonia (number 6 in 1990), nitric acid (number 13 in 1990), and ammonium nitrate (number 14 in 1990). All three of these compounds are used extensively in agriculture as synthetic fertilizers.

More than 80 percent of the ammonia produced, for example, goes into the production of synthetic fertilizers.

In addition to its agricultural role, nitric acid is an important raw material in the production of explosives. Trinitrotoluene (TNT), gunpowder, nitroglycerin, dynamite, and smokeless powder are all examples of the kind of explosives made from nitric acid. Slightly more than 5 percent of the nitric acid produced is also used in the synthesis of adipic acid and related compounds used in the manufacture of nylon.

Phosphorus

Phosphorus exists in three allotropic forms (physically or chemically different forms of the same substance): white, red, and black. The white form of phosphorus is a highly active, waxy solid that catches fire spontaneously when exposed to air. In contrast, red phosphorus is a reddish powder that is relatively inert (inactive). It does not catch fire unless exposed to an open flame. The melting point of phosphorus is 44°C (111°F), and its boiling point is 280°C (536°F). It is the eleventh most abundant element in Earth's crust.

Phosphorus always occurs in the form of a phosphate, a compound consisting of phosphorus, oxygen, and at least one more element. By far the most abundant source of phosphorus on Earth is a family of minerals known as the apatites. Apatites contain phosphorus, oxygen, calcium, and a halogen (chlorine, fluorine, bromine, or iodine). The state of Florida is the world's largest producer of phosphorus and is responsible for about a third of all the element produced in the world.

Phosphorus also occurs in all living organisms, most abundantly in bones, teeth, horn, and similar materials. It is found in all cells, however, in the form of compounds essential to the survival of all life. Like carbon and nitrogen, phosphorus is cycled through the environment. But since it has no common gaseous compounds, the phosphorus cycle occurs entirely within the solid and liquid (water) portions of Earth's crust.

Uses. About 95 percent of all the phosphorus used in industry goes to the production of phosphorus compounds. By far the most important of these is phosphoric acid, which accounts for about 83 percent of all phosphorus use in industry. A minor use is in the manufacture of safety matches.

Phosphoric acid. Phosphoric acid (H_3PO_4) typically ranks about number seven among the chemicals most widely produced in the United States. It is converted to a variety of forms, all of which are then used in the manufacture of synthetic fertilizer, accounting for about 85 percent of all the acid produced. Other applications of phosphoric acid include the pro-

Nitrogen family

duction of soaps and detergents, water treatment, the cleaning and rust-proofing of metals, the manufacture of gasoline additives, and the production of animal feeds.

At one time, large amounts of phosphoric acid were converted to a compound known as sodium tripolyphosphate (STPP). STPP, in turn, was used in the manufacture of synthetic detergents. When STPP is released to the environment, however, it serves as a primary nutrient for algae in bodies of water such as ponds and lakes. The growth of huge algal blooms in the 1970s and 1980s as a result of phosphate discharges eventually led to bans on the use of this compound in detergents. As a consequence, the compound is no longer commercially important.

Arsenic and antimony

Arsenic and antimony are both metalloids. That is, they behave at times like metals and at times like nonmetals. Arsenic is a silver-gray brittle metal that tarnishes when exposed to air. It exists in two allotropic forms: black and yellow. Its melting point is 817°C (1502°F) at 28 atmospheres of pressure, and its boiling point is 613°C (1135°F), at which temperature it sublimes (passes directly from the solid to the vapor state).

Antimony also occurs in two allotropic forms: black and yellow. It is a silver-white solid with a melting point of 630°C (1170°F) and a boiling point of 1635°C (2980°F). Both arsenic and antimony were identified before the birth of modern chemistry—at least as early as the fifteenth century.

Arsenic is a relatively uncommon element in Earth's crust, ranking number 51 in order of abundance. It is actually produced commercially from the flue dust obtained from copper and lead smelters (metals separated by melting) since it generally occurs in combination with these two elements.

Antimony is much less common in Earth's crust than is arsenic, ranking number 62 among the elements. It occurs most often as the mineral stibnite (antimony sulfide), from which it is obtained in a reaction with iron metal.

Uses. Arsenic is widely employed in the production of alloys (a mixture of two or more metals or a metal and a nonmetal) used in shot, batteries, cable covering, boiler tubes, and special kinds of solder (a melted metallic alloy used to join together other metallic surfaces). In a very pure form, it is an essential component of many electronic devices. Traditionally, compounds of arsenic have been used to kill rats and other pests, although it has largely been replaced for that purpose by other products.

Antimony is also a popular alloying element. Its alloys can be found in ball bearings, batteries, ammunition, solder, type metal, sheet pipe, and other applications. Its application in type metal reflects an especially interesting property: unlike most materials, antimony expands as it cools and solidifies from a liquid. Because of that fact, type metal poured into dies in the shape of letters expands as it cools to fill all parts of the die. Letters formed in this process have clear, sharp edges.

Bismuth

Bismuth is a typical silvery metal with an interesting reddish tinge to it. It has a melting point of 271°C (520°F) and a boiling point of 1560°C (2840°F). It is one of the rarest elements in Earth's crust, ranking 69 out of 75 elements for which estimates have been made. It occurs most commonly as the mineral bismite (bismuth oxide), bismuthinite (bismuth sulfide) and bismutite (bismuth oxycarbonate). Like arsenic and antimony, bismuth was identified as early as the fifteenth century by the pre-chemists known as the alchemists.

Nearly all of the bismuth produced commercially is used for one of two applications: in the production of alloys or other metallic products and in pharmaceuticals. Some of its most interesting alloys are those that melt at low temperatures and that can be used, for example, in automatic sprinkler systems. Compounds of bismuth are used to treat upset stomach, eczema (a skin disorder), and ulcers, and in the manufacture of face powders.

Noble gases

The noble gases are the six elements that make up Group 18 of the periodic table: helium (He), neon (Ne), argon (Ar), krypton (Kr), xenon (Xe), and radon (Rn). At one time, this family of elements was also known as the rare gases. Their present name comes from the fact that the six gases are highly unreactive; they appear almost "noble"—above interacting with other members of the periodic table. This lack of reactivity has also led to a second name by which they are sometimes known—the inert gases. (Inert means inactive.)

Abundance and production

As their former name suggests, the noble gases are rather uncommon on Earth. Collectively, they make up about 1 percent of Earth's

Noble gases

atmosphere. Most of the noble gases have been detected in small amounts in minerals found in Earth's crust and in meteorites. They are thought to have been released into the atmosphere long ago as by-products of the decay of radioactive elements in Earth's crust. (Radioactivity is the property that some elements have of spontaneously giving off energy in the form of particles or waves when their nuclei disintegrate.)

Of all the rare gases, argon is present in the greatest amount. It makes up about 0.9 percent by volume of Earth's atmosphere. The other noble gases are present in such small amounts that it is usually more convenient to express their concentrations in terms of parts per million (ppm). The concentrations of neon, helium, krypton, and xenon are, respectively, 18 ppm, 5 ppm, 1 ppm, and 0.09 ppm. For example, there are only 5 liters of helium in every million liters of air. By contrast, helium is much more abundant in the Sun, stars, and outer space. In fact, next to hydrogen, helium is the most abundant element in the universe. About 23 percent of all atoms found in the universe are helium atoms.

Radon is present in the atmosphere in only trace amounts. However, higher levels of radon have been measured in homes around the United States. Radon can be released from soils containing high concentrations of uranium, and they can be trapped in homes that have been weather sealed to make heating and cooling systems more efficient. Radon test-

Lead canisters used to store xenon for medical diagnostic purposes. *(Reproduced by permission of Photo Researchers, Inc.)*

ing kits are commercially available for testing the radon content of household air.

Most of the rare gases are obtained commercially from liquid air. As the temperature of liquid air is raised, the rare gases boil off from the mixture at specific temperatures and can be separated and purified. Although present in air, helium is obtained commercially from natural gas wells where it occurs in concentrations of between 1 and 7 percent of the natural gas. Most of the world's helium supplies come from wells located in Texas, Oklahoma, and Kansas. Radon is isolated as a product of the radioactive decay of radium compounds.

Properties

The noble gases are all colorless, odorless, and tasteless. They exist as monatomic gases, which means that their molecules consist of a single atom apiece. The boiling points of the noble gases increase in moving down the periodic table. Helium has the lowest boiling point of any element. It boils at 4.215 K ($-268.93°C$). It has no melting point because it cannot be frozen at any temperature.

The most important chemical property of the noble gases is their lack of reactivity. Helium, neon, and argon do not combine with any other elements to form compounds. It has been only in the last few decades that compounds of the other rare gases have been prepared. In 1962 English chemist Neil Bartlett (1932–) succeeded in preparing the first compound of a noble gas, a compound of xenon. The compound was xenon platinofluoride ($XePtF_6$). Since then, many xenon compounds containing mostly fluorine or oxygen atoms have also been prepared. Krypton and radon have also been combined with fluorine to form simple compounds. Because some noble gas compounds have powerful oxidizing properties, they have been used to synthesize other compounds.

The low reactivity of the noble gases can be explained by their electronic structure. The atoms of all six gases have outer energy levels containing eight electrons. Chemists believe that such arrangements are the most stable arrangements an atom can have. Because of these very stable arrangements, noble gas atoms have little or no tendency to gain or lose electrons, as they would have to do to take part in a chemical reaction.

Uses

As with all substances, the uses to which the noble gases are put reflect their physical and chemical properties. For example, helium's low density and inertness make it ideal for use in lighter-than-air craft such as balloons and dirigibles (zeppelins). Because of the element's very low boiling point, it has many applications in low-temperature research and technology. Divers breathe an artificial oxygen-helium mixture to prevent the formation of gas bubbles in the blood as they swim to the surface from great depths. Other uses for helium have been in supersonic wind tunnels, as a protective gas in growing silicon and germanium crystals and, together with neon, in the manufacture of gas lasers.

Neon is well known for its use in neon signs. Glass tubes of any shape can be filled with neon. When an electrical charge is passed through the tube, an orange-red glow is emitted. By contrast, ordinary incandescent lightbulbs are filled with argon. Because argon is so inert, it does not react with the hot metal filament and prolongs the bulb's life. Argon is also used to provide an inert atmosphere in welding and high-temperature metallurgical processes. By surrounding hot metals with inert argon, the metals are protected from potential oxidation by oxygen in the air.

Krypton and xenon also find commercial lighting applications. Krypton can be used in incandescent lightbulbs and in fluorescent lamps. Both are also employed in flashing stroboscopic lights that outline commercial airport runways. And because they emit a brilliant white light when electrified, they are used in photographic flash equipment. Due to the radioactive nature of radon, it has medical applications in radiotherapy.

[*See also* **Element, chemical; Periodic table**]

North America

North America, the world's third-largest continent, encompasses an area of about 9,400,000 square miles (24,346,000 square kilometers). This landmass is occupied by the present-day countries of Canada, the United States, Mexico, Guatemala, Belize, El Salvador, Honduras, Nicaragua, Costa Rica, and Panama. Also included in the North American continent are Greenland, an island landmass northeast of Canada, and the islands of the Caribbean, many of which are independent republics.

North America is bounded on the north by the Arctic Ocean, on the west by the Bering Sea and the Pacific Ocean, on the south by the South American continent, and on the east by the Gulf of Mexico and Atlantic Ocean.

The North American continent contains almost every type of landform present on Earth: mountains, forests, plateaus, rivers, valleys, plains, deserts, and tundra. It also features every type of climatic zone found

Opposite Page: North America. *(Reproduced by permission of The Gale Group.)*

North America

North America

on Earth, from polar conditions in Greenland to tropical rain forests in the countries of Central America. Much of the continent, however, is subject to a temperate climate, resulting in favorable farming and living conditions.

The highest point on the continent is Mount McKinley in Alaska, standing 20,320 feet (6,194 meters) in height. Badwater, in the south-central part of Death Valley in California, is the continent's lowest point, at 282 feet (86 meters) below sea level.

Coast of Lake Michigan at Indiana Dunes National Lakeshore, Indiana. *(Reproduced by permission of the National Parks Service.)*

Rivers and lakes

The North American continent contains the world's greatest inland waterway system. The Mississippi River rises in northern Minnesota and flows 2,348 miles (3,778 kilometers) down the center of the United States to the Gulf of Mexico. The Missouri River, formed by the junction of three rivers in southern Montana, runs 2,466 miles (3,968 kilometers) before it joins the Mississippi just north of St. Louis, Missouri. The Ohio River, formed by the union of two rivers at Pittsburgh, Pennsylvania, flows 975 miles (1,569 kilometers) before emptying into the Mississippi at Cairo, Illinois. The Mississippi, with all of its tributaries, drains 1,234,700 square miles (3,197,900 square kilometers) from all or part of 31 states in the United States. From the provinces of Alberta and Saskatchewan in Canada, the Mississippi drains about 13,000 square miles (33,670 square kilometers).

Other chief rivers in North America include the Yukon (Alaska and Canada); Mackenzie, Nelson, and Saskatchewan (Canada); Columbia and St. Lawrence (Canada and U.S.); Colorado, Delaware, and Susquehanna (U.S.); and Rio Grande (U.S. and Mexico).

North America contains more lakes than any other continent. Dominant lakes include Great Bear, Great Slave, and Winnipeg (Canada); the Great Lakes (Canada and U.S.); Great Salt Lake (U.S.); Chapala (Mexico); and Nicaragua (Nicaragua). The Great Lakes, a chain of five lakes, are Superior, Michigan, Huron, Erie, and Ontario. Superior, northernmost and westernmost of the five, is the largest lake in North America and the largest body of freshwater in the world. Stretching 350 miles (560 kilometers) long, the lake covers about 31,820 square miles (82,410 square kilometers). It has a maximum depth of 1,302 feet (397 meters).

Geographical regions

Geologists divide the North American continent into a number of geographical regions. The five main regions are the Canadian Shield, the Appalachian System, the Coastal Plain, the Central Lowlands, and the North American Cordillera (pronounced kor-dee-YARE-ah; a complex group of mountain ranges, systems, and chains).

Canadian Shield. The Canadian Shield is a U-shaped plateau region of very old, very hard rocks. It was the first part of North America to be elevated above sea level, and became the central core around which geological forces built the continent. It is sometimes called the Laurentian Plateau. It extends north from the Great Lakes to the Arctic Ocean, covering more than half of Canada and including Greenland. Hudson Bay and Foxe Basin in Canada mark the center of the region, submerged by the weight of glaciers of the most recent ice age some 11,000 years ago. Mountains ranges ring the outer edges of this geological structure. In the United States, the Adirondack Mountains and the Superior Highlands are part of the Shield.

The southern part of the Canadian Shield is covered by rich forests, while the northern part is tundra (rolling, treeless plains). The region is rich in minerals, including cobalt, copper, gold, iron, nickel, uranium, and zinc.

Appalachian System. The Appalachian Mountains extend about 1,600 miles (2,570 kilometers) southwest from Newfoundland to Alabama. They are a geologically old mountain system. Formed over 300 million years ago, the Appalachians have eroded greatly since then. Most of the system's ridges are 1,200 to 2,400 feet (360 to 730 meters) in height. Only a few peaks rise above 6,000 feet (1,800 meters). The system's highest peak, Mount Mitchell, rises 6,684 feet (2,037 meters) above sea level.

The main ranges in the system are the White Mountains (New Hampshire), Green Mountains (Vermont), Catskill Mountains (New York), Allegheny Mountains (Pennsylvania), Great Smoky Mountains (North Carolina and Tennessee), Blue Ridge Mountains (Pennsylvania to Georgia), and the Cumberland Mountains (West Virginia to Alabama).

Much mineral wealth is found throughout the Appalachian System, including coal, iron, lead, zinc, and bauxite. Other mineral resources such as petroleum and natural gas are also prevalent.

Coastal Plain. The Coastal Plain is a belt of lowlands that extends from southern New England to Mexico's Yucatan Peninsula, flanking the Atlantic Ocean and the Gulf of Mexico. This geological area was the last part added to the North American continent. Much of the plain lies underwater along the northern Atlantic Coast, forming rich fishing banks.

The southern portion of the plain, from Florida along the Gulf shore of Louisiana and Texas into Mexico, holds large deposits of phosphate, salt, and sulfur. Extensive oil and natural gas fields also line this area.

Central Lowlands. The Central Lowlands extend down the center of the continent from the Mackenzie Valley in the Northwest Territories in Canada to the Coastal Plain in the Gulf of Mexico. These lowlands circle the Canadian Shield. Included in this extensive region are the Great Plains in the west and the lowlands of the Ohio-Great Lakes-Mississippi area in the east. The great North American rivers are contained in this region, making the surrounding soil fertile for farming. The world's richest sources of coal, oil, and natural gas are also found here.

North American Cordillera. The North American Cordillera is a complex group of geologically young mountains that extend along the western edge of the North American continent. The eastern section of the Cordillera is marked by the Rocky Mountains. They extend more than 3,000 miles (4,800 kilometers) from northwest Alaska to central New Mexico. The highest peak in the Rockies is Mount Elbert in Colorado at 14,431 feet (4,399 meters) in height. The highest peak in the Canadian Rockies is Mount Robson in eastern British Columbia, rising 12,972 feet (3,954 meters). The ridge of the Rocky Mountains is known as the Continental Divide, the "backbone" of the continent that separates the rivers draining to the Arctic and Atlantic Oceans from those draining to the Pacific Ocean.

Snow-covered Mt. Sopris on the Crystal River near Aspen, Colorado. *(Reproduced by permission of Photo Researchers, Inc.)*

North America

The Rockies may be divided into three sections: northern, central, and southern. The Northern Rockies, which rise to great elevations, begin in northern Alaska and extend down into Montana. From here, the Central Rockies extend down into Colorado. A high, vast plateau separates the Central Rockies from the Southern Rockies. Known as the Wyoming Basin, it varies in elevation from 7,000 to 8,000 feet (2,100 to 2,400 meters). The Southern Rockies contain the highest peaks in the entire system—many exceed 14,000 feet (4,300 meters) in height.

West of the Rockies lies a series of plateaus and basins. These include the Yukon Plateau, the uplands in central British Columbia, the Snake River Plain, the Great Basin, and the Colorado Plateau. The Great Basin, an elevated region between the Wasatch and Sierra Nevada Mountains, includes the Great Salt Lake, the Great Salt Lake and Mojave deserts, and Death Valley.

The western edge of North America is marked by two mountain ranges: the Cascade and Coast ranges. The Cascade Range extends about 700 miles (1,130 kilometers) from British Columbia through Washington and Oregon into northeast California. Many of the range's peaks are volcanic in origin. The highest peak is Mount Rainier in Washington, standing 14,410 feet (4,390 meters) in height. North of the Cascades are the Coast Mountains, which extend about 1,000 miles (1,610 kilometers) north from British Columbia into southeast Alaska. Here they are met by the Alaska Range, which extends in a great arc through south-central Alaska. This range features the highest peaks in North America, including Mount McKinley.

South of the Cascades are the Sierra Nevada Mountains, extending about 400 miles (640 kilometers) through eastern California. The Sierras, noted along with the Cascades for their beauty, contain Mount Whitney. At 14,494 feet (4,418 meters) tall, it is the highest peak in the contiguous United States (the 48 connected states).

The Coast Ranges are a series of mountain ranges along North America's Pacific coast. They extend from southeast Alaska to Baja California. The ranges include the St. Elias Mountains (Alaska and Canada); Olympic Mountains (Washington); Coast Ranges (Oregon); Klamath Mountains, Coast Ranges, and Los Angeles Ranges (California); and the Peninsular Range (Baja California). Peaks in the entire Coast Ranges extend from 2,000 to 20,000 feet (610 to 6,100 meters) in height.

In Mexico, the chief mountain system is the Sierra Madre, composed of the Sierra Madre Occidental, the Sierra Madre Oriental, and the Sierra Madre del Sur. The Sierra Madre Occidental begins just south of the Rio Grande River and runs about 700 miles (1,130 kilometers) parallel to the

Gulf of Mexico. The Occidental contains the highest peak in the Sierra Madre system, Pico de Orizaba, which rises to 18,700 feet (5,700 meters). Orizaba is also considered a part of the Cordillera de Anahuac, an east-west running belt of lofty volcanoes just south of Mexico City. In addition to Orizaba, this belt contains the volcanic peaks Popocatepetl and Ixtacihuatl. The belt connects the Occidental range to the Sierra Madre Oriental, which runs south from Arizona parallel to the Pacific coast for about 1,000 miles (1,610 kilometers). The Sierra Madre del Sur is a broken mass of uptilted mountains along the Pacific coast in southern Mexico. It forms the natural harbor of Acapulco.

Nova

The word *nova*, Latin for "new," was assigned by ancient astronomers to any bright star that suddenly appeared in the sky. A nova occurs when

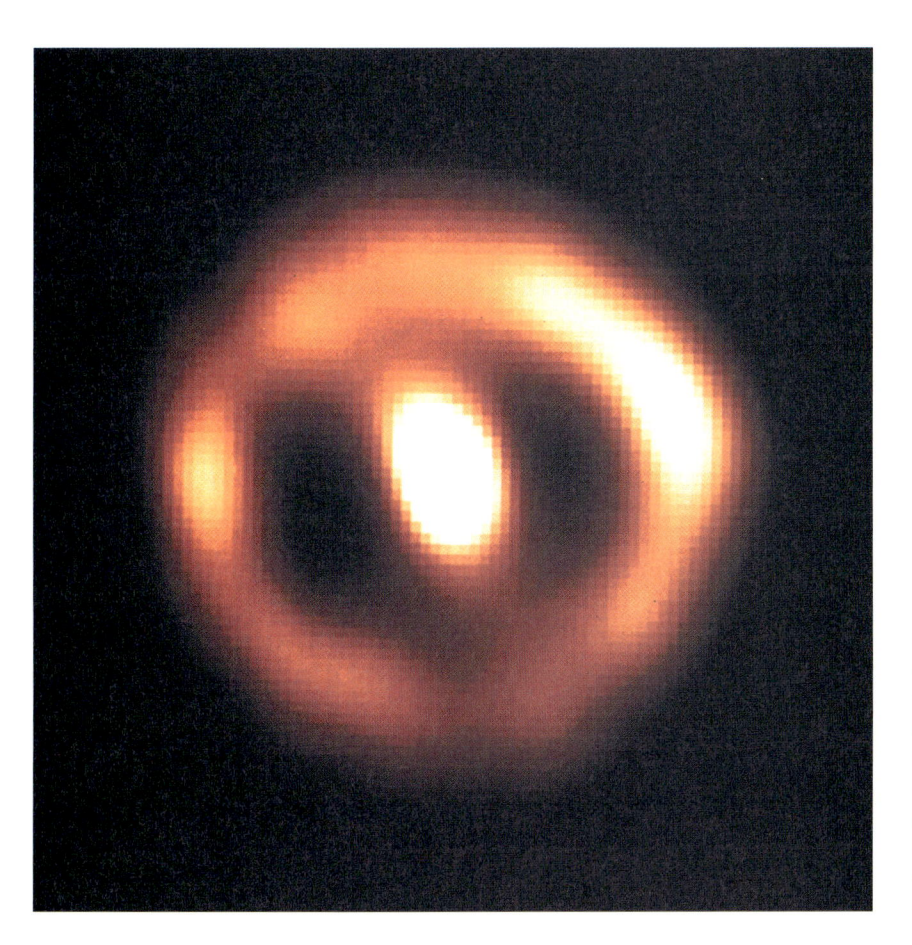

Ultraviolet image of Nova Cygni 1992. On February 19, 1992, this nova was formed by an explosion triggered by the transfer of gases to the white dwarf from its companion star. *(Reproduced by permission of National Aeronautics and Space Administration.)*

Nova

> **Words to Know**
>
> **Binary star:** Pair of stars in a single system that orbit each other, bound together by their mutual gravities.
>
> **Red giant:** A medium-sized star in a late stage of its evolution. It is relatively cool and has a diameter that is perhaps 100 times its original size.
>
> **White dwarf:** The cooling, shrunken core remaining after a medium-sized star ceases to burn.

one member of a binary star system temporarily becomes brighter. Most often the brighter star is a shrunken white dwarf, the cooling, shrunken core remaining after a medium-sized star (like our sun) ceases to burn. Its partner is a large star, such as a red giant, a medium-sized star in a late stage of its evolution, expanding and cooling.

As the companion star expands, it loses some of its matter—mostly hydrogen—to the strong gravitational pull of the white dwarf. After a time, enough matter collects in a thin, dense, hot layer on the surface of the white dwarf to initiate nuclear fusion reactions. The hydrogen on the white dwarf's surface burns away, and while it does so, the white dwarf glows brightly. This is a nova. After reaching its peak brightness, it slowly fades over a period of days or weeks.

The transfer of matter does not stop after a nova explodes, but begins anew. The length of time between nova outbursts can range from several dozen to thousands of years, depending on how fast the companion star loses matter to the white dwarf.

A nova should not be confused with a supernova, which is the massive explosion of a relatively large star. A nova is much more common than a supernova, and it does not release nearly as much energy. Because novae (plural of nova) occur more often, they can change the way constellations in the night sky appear. For example, in December 1999, a bright, naked-eye nova appeared in the constellation Aquila, the Eagle. At its maximum, the nova was as bright as many of the stars in Aquila. For a few days at least, viewers were treated to the spectacle of a truly "new star" in an otherwise familiar constellation.

[*See also* **Binary star; Star; Supernova; White dwarf**]

Nuclear fission

Nuclear fission is a process in which the nucleus of a heavy atom is broken apart into two or more smaller nuclei. The reaction was first discovered in the late 1930s when a target of uranium metal was bombarded with neutrons. Uranium nuclei broke into two smaller nuclei of roughly equal size with the emission of very large amounts of energy. Some scientists immediately recognized the potential of the nuclear fission reaction for the production of bombs and other types of weapons as well as for the generation of power for peacetime uses.

History

The fission reaction was discovered accidentally in 1938 by two German physicists, Otto Hahn (1879–1968) and Fritz Strassmann (1902–1980). Hahn and Strassmann had been doing a series of experiments in which they used neutrons to bombard various elements. When they bombarded copper, for example, a radioactive form of copper was produced. Other elements became radioactive in the same way.

Their work with uranium, however, produced entirely different results. In fact, the results were so unexpected that Hahn and Strassmann were unable to offer a satisfactory explanation for what they observed. That explanation was provided, instead, by German physicist Lise Meitner (1878–1968) and her nephew Otto Frisch (1904–1979). Meitner was a longtime colleague of Hahn who had left Germany due to anti-Jewish persecution.

In most nuclear reactions, an atom changes from a stable form to a radioactive form, or it changes to a slightly heavier or a slightly lighter atom. Copper (element number 29), for example, might change from a stable form to a radioactive form or to zinc (element number 30) or nickel (element number 28). Such reactions were already familiar to nuclear scientists.

What Hahn and Strassmann had seen—and what they had failed to recognize—was a much more dramatic nuclear change. An atom of uranium (element number 92), when struck by a neutron, broke into two much smaller elements such as krypton (element number 36) and barium (element number 56). The reaction was given the name nuclear fission because of its similarity to the process by which a cell breaks into two parts during the process of cellular fission.

Putting nuclear fission to work

In every nuclear fission, three kinds of products are formed. The first product consists of the smaller nuclei produced during fission. These

Nuclear fission

> ## Words to Know
>
> **Chain reaction:** A reaction in which a substance needed to initiate a reaction is also produced as the result of that reaction.
>
> **Fission products:** The isotopes formed as the result of a nuclear fission reaction.
>
> **Fission weapon:** A bomb or other type of military weapon whose power is derived from a nuclear fission reaction.
>
> **Isotopes:** Two or more forms of an element that have the same chemical properties but that differ in mass because of differences in the number of neutrons in their nuclei.
>
> **Manhattan Project:** A research project of the United States government created to develop and produce the world's first atomic bomb.
>
> **Mass:** A measure of the amount of matter in a body.
>
> **Neutron:** A subatomic particle with a mass about equal to that of a hydrogen atom but with no electric charge.
>
> **Nuclear reactor:** Any device for controlling the release of nuclear power so that it can be used for constructive purposes.
>
> **Radioactivity:** The property possessed by some elements of spontaneously emitting energy in the form of particles or waves by disintegration of their atomic nuclei.
>
> **Radioactive isotope:** An isotope that spontaneously breaks down into another isotope with the release of some form of radiation.
>
> **Subatomic particle:** Basic unit of matter and energy (proton, neutron, electron, neutrino, and positron) smaller than an atom.

nuclei, like krypton and barium in the example mentioned above, are called fission products. Fission products are of interest for many reasons, one of which is that they are always radioactive. That is, any time a fission reaction takes place, radioactive materials are formed as by-products of the reaction.

The second product of a fission reaction is energy. A tiny amount of matter in the original uranium atom is changed into energy. In the early 1900s, German-born American physicist Albert Einstein (1879–1955) had showed how matter and energy can be considered two forms of the same phenomenon. The mathematical equation that represents this relationship, $E = mc^2$, has become one of the most famous scientific formulas in the world. The formula says that the amount of energy (E) that can be obtained from a certain amount of matter (m) can be found by multiplying that amount of matter by the square of the speed of light (c^2). The square of the speed of light is a very large number, equal to about 9×10^{20} meters per second, or 900,000,000,000,000,000 meters per second. Thus, if even a very small amount of matter is converted to energy, the amount of energy obtained is very large. It is this availability of huge amounts of energy that originally made the fission reaction so interesting to both scientists and nonscientists.

The third product formed in any fission reaction is neutrons. The significance of this point can be seen if you recall that a fission reaction is initiated when a neutron strikes a uranium nucleus or other large nucleus. Thus, the particle needed to *originate* a fission reaction is also *produced* as a result of the reaction.

Chain reactions. Imagine a chunk of uranium metal consisting of trillions upon trillions of uranium atoms. Then imagine that a single neutron is fired into the chunk of uranium, as shown in the accompanying figure of a nuclear chain reaction. If that neutron strikes a uranium nucleus, it can cause a fission reaction in which two fission products and two neutrons are formed. Each of these two neutrons, in turn, has the potential for causing the fission of two other uranium nuclei. Two neutrons produced in *each* of those two reactions can then cause fission in four uranium nuclei. And so on.

In actual practice, this series of reactions, called a chain reaction, takes place very rapidly. Millions of fission reactions can occur in much less than a second. Since energy is produced during each reaction, the total amount of energy produced throughout the whole chunk of uranium metal is very large indeed.

The first atomic bomb

Perhaps you can see why some scientists immediately saw fission as a way of making very powerful bombs. All you have to do is to find a large enough chunk of uranium metal, bombard the uranium with neutrons, and get out of the way. Fission reactions occur trillions of times over again in a short period of time, huge amounts of energy are released, and the uranium blows apart, destroying everything in its path. Pictures of actual atomic bomb blasts vividly illustrate the power of fission reactions.

But the pathway from the Hahn/Strassmann/Meitner/Frisch discovery to an actual bomb was a long and difficult one. A great many technical problems had to be solved in order to produce a bomb that worked on the principle of nuclear fission. One of the most difficult of those problems involved the separation of uranium-238 from uranium-235.

Naturally occurring uranium consists of two isotopes: uranium-238 and uranium-235. Isotopes are two forms of the same element that have the same chemical properties but different masses. The difference between these two isotopes of uranium is that uranium-235 nuclei will undergo nuclear fission, but those of uranium-238 will not. That problem is compounded by the fact that uranium-238 is much more abundant in nature than is uranium-235. For every 1,000 atoms of uranium found in Earth's crust, 993 are atoms of uranium-238 and only 7 are atoms of uranium-235. One of the biggest problems in making fission weapons a reality, then, was finding a way to separate uranium-235 (which *could* be used to make bombs) from uranium-238 (which *could not,* and thus just got in the way).

The Manhattan Project. A year into World War II (1939–45), a number of scientists had come to the conclusion that the United States would have to try building a fission bomb. They believed that Nazi Germany would soon be able to do so, and the free world could not survive unless it, too, developed fission weapons technology.

Thus, in 1942, President Franklin D. Roosevelt authorized the creation of one of the largest and most secret research operations ever devised. The project was given the code name Manhattan Engineering District, and its task was to build the world's first fission (atomic) bomb. That story is a long and fascinating one, a testimony to the technological miracles that can be produced under the pressures of war. The project reached its goal on July 16, 1945, in a remote part of the New Mexico desert, where the first atomic bomb was tested. Less than a month later, the first fission bomb was actually used in war. It was dropped on the Japanese city of Hiroshima, destroying the city and killing over 80,000 people. Three days later, a second bomb was dropped on Nagasaki, with similar results. For all the horror they caused, the bombs seemed to have achieved their objective. The Japanese leaders appealed for peace only three days after the Nagasaki event. (Critics, however, charge that the end of the war was in sight and that the Japanese would have surrendered without the use of a devastating nuclear weapon.)

Nuclear fission in peacetime

The world first learned about the power of nuclear fission in the form of terribly destructive weapons, the atomic bombs. But scientists had long known that the same energy released in a nuclear weapon could be harnessed for peacetime uses. The task is considerably more difficult, however. In a nuclear weapon, a chain reaction is initiated—energy is produced and released directly to the environment. In a nuclear power reactor, however, some means must be used to control the energy produced in the chain reaction.

The control of nuclear fission energy was actually achieved before the production of the first atomic bomb. In 1942, a Manhattan Project research team under the direction of Italian physicist Enrico Fermi (1901–1954) designed and built the first nuclear reactor. A nuclear reactor is a device for obtaining the controlled release of nuclear energy. The reactor had actually been built as a research instrument to learn more about nuclear fission (as a step in building the atomic bomb).

After the war, the principles of Fermi's nuclear reactor were used to construct the world's first nuclear power plants. These plants use the

Nuclear fission

A nuclear chain reaction: the uninterrupted fissioning of ever-increasing numbers of uranium-235 atoms. *(Reproduced by permission of The Gale Group.)*

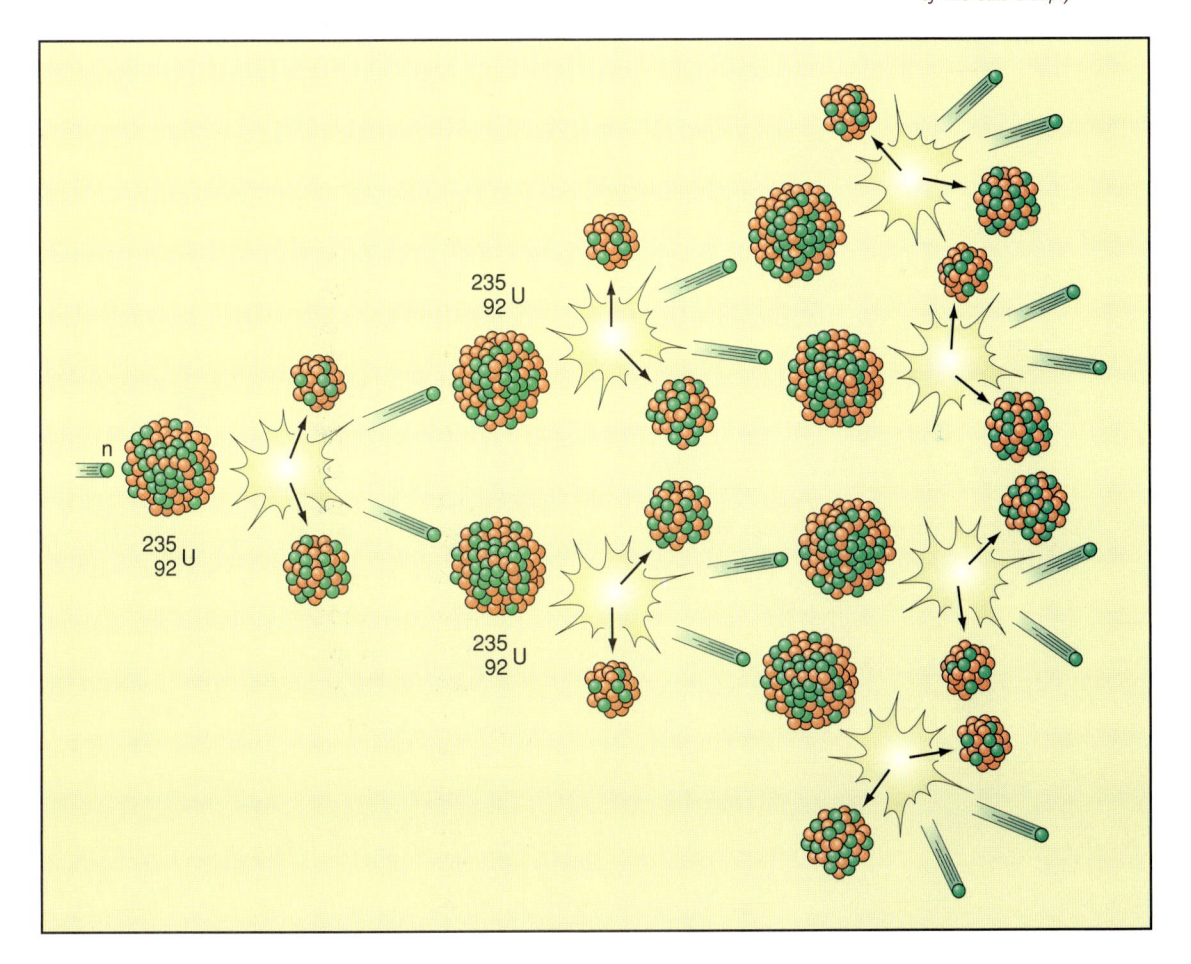

energy released by nuclear fission to heat water in boilers. The steam that is produced is then used to operate turbines and electrical generators. The first of these nuclear power plants was constructed in Shippingport, Pennsylvania, in 1957. In the following three decades, over 100 more nuclear power plants were built in every part of the United States, and at least as many more were constructed throughout the world.

By the dawn of the 1990s, however, progress in nuclear power production had essentially come to a stop in the United States. Questions about the safety of nuclear power plants had not been answered to the satisfaction of most Americans, and, as a result, no new nuclear plants have been built in the United States since the mid-1980s.

Despite these concerns, nuclear power plants continue to supply a good portion of the nation's electricity. Since 1976, nuclear electrical generation has more than tripled. At the beginning of the twenty-first century, 104 commercial nuclear power reactors in 31 states accounted for about 22 percent of the total electricity generated in the country. Combined, coal and nuclear sources produce 78 percent of the nation's electricity.

[*See also* **Nuclear fusion; Nuclear power; Nuclear weapons**]

Nuclear fusion

Nuclear fusion is the process by which two light atomic nuclei combine to form one heavier atomic nucleus. As an example, a proton and a neutron can be made to combine with each other to form a single particle called a deuteron. In general, the mass of the heavier product nucleus (the deuteron, for example) is less than the total mass of the two lighter nuclei (the proton and the neutron).

The mass that "disappears" during fusion is actually converted into energy. The amount of energy (E) produced in such a reaction can be calculated using Einstein's formula for the equivalence of mass and energy: $E = mc^2$. This formula says even when the amount of mass (m) that disappears is very small, the amount of energy produced is very large. The reason is that the value of c^2 (the speed of light squared) is very large, approximately 900,000,000,000,000,000,000 meters per second.

Naturally occurring fusion reactions

Scientists have long suspected that nuclear fusion reactions are common in the universe. The factual basis for such beliefs is that stars consist primarily of hydrogen gas. Over time, however, hydrogen gas is used

Words to Know

Cold fusion: A form of fusion that some researchers believe can occur at or near room temperatures as the result of the combination of deuterons during the electrolysis of water.

Deuteron: The nucleus of the deuterium atom, consisting of one proton combined with one neutron.

Electrolysis: The process by which an electrical current causes a chemical change, usually the breakdown of some substance.

Isotopes: Two or more forms of an element that have the same chemical properties but that differ in mass because of differences in the number of neutrons in their nuclei.

Neutron: A subatomic particle with a mass of about one atomic mass unit and no electrical charge.

Nuclear fission: A nuclear reaction in which one large atomic nucleus breaks apart into at least two smaller particles.

Nucleus: The core of an atom consisting of one or more protons and, usually, one or more neutrons.

Plasma: A form of matter that consists of positively charged particles and electrons completely independent of each other.

Proton: A subatomic particle with a mass of about one atomic mass unit and a single positive charge.

Subatomic particle: Basic unit of matter and energy (proton, neutron, electron, neutrino, and positron) smaller than an atom.

Thermonuclear reaction: A nuclear reaction that takes place only at very high temperatures, usually on the order of a few million degrees.

---

up in stars, and helium gas is produced. One way to explain this phenomenon is to assume that hydrogen nuclei in the core of stars fuse with each other to form the nuclei of helium atoms. That is:

$$4 \text{ hydrogen nuclei} \rightarrow \text{fuse} \rightarrow 1 \text{ helium nucleus}$$

Over the past half century, a number of theories have been suggested as to how such fusion reactions might occur. One problem that must be resolved in such theories is the problem of electrostatic repulsion. Electrostatic repulsion is the force that tends to drive two particles with the same electric charge away from each other.

Nuclear fusion

The nucleus of a hydrogen atom is a single proton, a positively charged particle. If fusion is to occur, two protons must combine with each other to form a single particle:

$$p^+ + p^+ \rightarrow \text{combined particle}$$

But forcing two like-charged particles together requires a lot of energy. Where do stars get that energy?

Thermonuclear reactions

The answer to that question has many parts, but one part involves heat. If you raise the temperature of hydrogen gas, hydrogen atoms move faster and faster. They collide with each other with more and more energy. Eventually, they may collide in such a way that two protons will combine with (fuse with) each other. Reactions that require huge amounts of energy in order to occur are called thermonuclear reactions: *thermo-* means "heat" and *-nuclear* refers to the nuclei involved in such reactions.

The amount of heat needed to cause such reactions is truly astounding. It may require temperatures from a few millions to a few hundred millions of degrees Celsius. Such temperatures are usually unknown on Earth, although they are not uncommon at the center of stars.

Scientists now believe that fusion reactions are the means by which stars generate their energy. In these reactions, hydrogen is first converted to helium, with the release of large amounts of energy. At some point, no more hydrogen is available for fusion reactions, a star collapses, it heats up, and new fusion reactions begin. In the next stage of fusion reactions, helium nuclei may combine to form carbon nuclei. This stage of reactions requires higher temperatures but releases more energy. When no more helium remains for fusion reactions, yet another sequence of reactions begin. This time, carbon nuclei might be fused in the production of oxygen or neon nuclei. Again, more energy is required for such reactions, and more energy is released.

The end result of this sequence of fusion reactions is that stars heat up to temperatures they can no longer withstand. They explode as novas or supernovas, releasing to the universe the elements they have been creating in their cores.

Fusion reactions on Earth

Dreams of harnessing fusion power for human use developed alongside similar dreams for harnessing fission power. The first step in the realization of those dreams—creating a fusion bomb—was relatively simple, requiring a large batch of hydrogen (like the hydrogen in a star) and a source of heat that would raise the temperature of the hydrogen to a few million degrees Celsius.

Nuclear fusion

Encapsulating the hydrogen was the easy part. A large container (the bomb casing) was built and filled with as much hydrogen as possible, probably in the form of liquid hydrogen. Obtaining the high temperature was more difficult. In general, there is no way to produce a temperature of 10,000,000°C on Earth. The only practical way to do so is to set off a fission (atomic) bomb. For a few moments after a fission bomb explodes, it produces temperatures in this range.

All that was needed to make a fusion bomb, then, was to pack a fission bomb at the center of the hydrogen-filled casing of the fusion bomb. When the fission bomb exploded, a temperature of a few million degrees Celsius would be produced, and fusion would begin within the hydrogen. As fusion proceeded, even greater amounts of energy would be produced, resulting in a bomb that was many times more powerful than the fission bomb itself.

For comparison, the fission bomb dropped on Hiroshima, Japan, in August of 1945 was given a power rating of about 20 kilotons. The measure 20 kilotons means that the bomb released as much energy as 20,000 tons of TNT, one of the most powerful chemical explosives known. In contrast, the first fusion (hydrogen) bomb tested had a power rating of 5 megatons, that is, the equivalent of 5 million tons of TNT.

Peaceful applications of nuclear fusion

As with nuclear fission, scientists were also very much committed to finding peaceful uses of nuclear fusion. The problems to be solved in controlling nuclear fusion reactions have, however, been enormous. The most obvious challenge is simply to find a way to "hold" the nuclear fusion reaction in place as it occurs. One cannot build a machine made out of metal, plastic, glass, or any other common kind of material. At the temperatures at which fusion occurs, any one of these materials would vaporize instantly. So how can the nuclear fusion reaction be contained?

One of the methods that has been tried is called magnetic confinement. To understand this technique, imagine that a mixture of hydrogen isotopes has been heated to a very high temperature. At a sufficiently high temperature, the nature of the mixture begins to change. Atoms totally lose their electrons, and the mixture consists of a swirling mass of

Nuclear fusion

positively charged nuclei and negatively charged electrons. Such a mixture is known as a plasma.

One way to control that plasma is with a magnetic field, which can be designed so that the swirling hot mass of plasma within the field is held in any kind of shape. The best known example of this approach is a doughnut-shaped Russian machine known as a *tokamak*. In the *tokamak*, two powerful electromagnets create fields that are so strong they can hold a hot plasma in place as readily as a person can hold an orange in his or her hand.

Tokamak 15, a nuclear fusion research reactor at the Kurchatov Institute in Moscow. The ring shape of the reactor is the design most favored by nuclear fusion researchers. The ring contains a plasma mixture of deuterium and tritium that is surrounded by powerful magnets that enclose the plasma with their fields and keep it away from the walls of the reactor vessel. At sufficiently high temperatures, the deuterium and tritium nuclei fuse, creating helium and energetic neutrons. It is these neutrons that carry the energy of the reactor. *(Reproduced by permission of Photo Researchers, Inc.)*

The technique, then, is to heat the hydrogen isotopes to higher and higher temperatures while containing them within a confined space by means of the magnetic fields. At some critical temperatures, nuclear fusion will begin to occur. At that point, the *tokamak* is producing energy by means of fusion while the fuel is being held in suspension by the magnetic field.

Hope for the future

Research on controlled fusion power has now been going on for a half century with somewhat disappointing results. Some experts argue that no method will ever be found for making fusion power by a method that humans can afford. The amount of energy produced by fusion, they say, will always be less than the amount of energy put into the process in the first place. Other scientists disagree. They believe that success may be soon in coming, and it is just a matter of finding solutions to the many technical problems surrounding the production of fusion power.

Cold fusion

The scientific world was astonished in March of 1989 when two electrochemists, Stanley Pons and Martin Fleischmann, reported that they had obtained evidence of the occurrence of nuclear fusion at room temperatures. Pons and Fleischmann passed an electric current through a form of water known as heavy water, or deuterium oxide. In the process, they reported fusion of deuterons had occurred. A deuteron is a particle consisting of a proton combined with a neutron. If such an observation could have been confirmed by other scientists, it would have been truly revolutionary: it would have meant that energy could be obtained from fusion reactions at moderate temperatures rather than at temperatures of millions of degrees.

The Pons-Fleischmann discovery was the subject of immediate and intense study by other scientists around the world. It soon became apparent, however, that evidence for cold fusion could not be obtained by other researchers with any degree of consistency. A number of alternative explanations were developed by scientists for the fusion results that Pons and Fleischmann believed they had obtained. Today, some scientists are still convinced that Pons and Fleischmann made a real and important breakthrough in the area of fusion research. Most researchers, however, attribute the results they reported to other events that occurred during the electrolysis of the heavy water.

[*See also* **Nuclear fission; Nuclear power; Nuclear weapons**]

Nuclear medicine

Nuclear medicine is a special field of medicine in which radioactive materials are used to conduct medical research and to diagnose (detect) and treat medical disorders. The radioactive materials used are generally called radionuclides, meaning a form of an element that is radioactive.

Diagnosis

Radionuclides are powerful tools for diagnosing medical disorders for three reasons. First, many chemical elements tend to concentrate in one part of the body or another. As an example, nearly all of the iodine that humans consume in their diets goes to the thyroid gland. There it is used to produce hormones that control the rate at which the body functions.

Second, the radioactive form of an element behaves biologically in exactly the same way that a nonradioactive form of the element behaves. When a person ingests (takes into the body) the element iodine, for example, it makes no difference whether the iodine occurs in a radioactive or nonradioactive form. In either case, it tends to concentrate in the thyroid gland.

Third, any radioactive material spontaneously decays, breaking down into some other form with the emission of radiation. That radiation can be detected by simple, well-known means. When radioactive iodine enters the body, for example, its progress through the body can be followed with a Geiger counter or some other detection instrument. Such instruments pick up the radiation given off by the radionuclide and make a sound, cause a light to flash, or record the radiation in some other way.

If a physician suspects that a patient may have a disease of the thyroid gland, that patient may be given a solution to drink that contains radioactive iodine. The radioactive iodine passes through the body and into the thyroid gland. Its presence in the gland can be detected by means of a special device. The physician knows what the behavior of a normal thyroid gland is from previous studies; the behavior of this particular patient's thyroid gland can then be compared to that of a normal gland. The test therefore allows the physician to determine whether the patient's thyroid is functioning normally.

Treatment

Radionuclides can also be used to treat medical disorders because of the radiation they emit. Radiation has a tendency to kill cells. Under

Words to Know

Diagnosis: Any attempt to identify a disease or other medical disorder.

Isotopes: Two or more forms of an element that have the same chemical properties but that differ in mass because of differences in the number of neutrons in their nuclei.

Radioactivity: The property possessed by some elements of spontaneously emitting energy in the form of particles or waves by disintegration of their atomic nuclei.

Radioactive decay: The process by which an isotope breaks down to form a different isotope, with the release of radiation.

Radioactive isotope: A form of an element that gives off radiation and changes into another isotope.

Radionuclide: A radioactive isotope.

many circumstances, that tendency can be a dangerous side effect: anyone exposed to high levels of radiation may become ill and can even die. But the cell-killing potential of radiation also has its advantages. A major difference between cancer cells and normal cells, for example, is that the former grow much more rapidly than the latter. For this reason, radiation can be used to destroy the cells responsible for a patient's cancer.

A radionuclide frequently used for this purpose is cobalt-60. It can be used as follows. A patient with cancer lies on a bed surrounded by a large machine that contains a sample of cobalt-60. The machine is then rotated in such a way around the patient's body that the radiation released by the sample is focused directly on the cancer. That radiation kills cancer cells and, to a lesser extent, some healthy cells too. If the treatment is successful, the cancer may be destroyed, producing only modest harm to the patient's healthy cells. That "modest harm" may occur in the form of nausea, vomiting, loss of hair, and other symptoms of radiation sickness that accompany radiation treatment.

Radioactive isotopes can be used in other ways for the treatment of medical disorders. For example, suppose that a patient has a tumor on his or her thyroid. One way of treating that tumor might be to give the patient a dose of radioactive iodine. In this case, the purpose of the iodine

Some Diagnostic Radionuclides Used in Medicine

Radionuclide	Use
Chromium-51	Volume of blood and of red blood cells
Cobalt-58	Uptake (absorption) of vitamin B_{12}
Gallium-67	Detection of tumors and abscesses
Iodine-123	Thyroid studies
Iron-59	Rate of formation/lifetime of red blood cells
Sodium-24	Studies of the circulatory system
Thallium-201	Studies of the heart
Technetium-99	Many kinds of diagnostic studies

is not to diagnose a disorder, but to treat it. When the iodine travels to the thyroid, the radiation it gives off may attack the tumor cells present there, killing those cells and thereby destroying the patient's tumor.

[*See also* **Isotope**]

Nuclear power

Nuclear power is any method of doing work that makes use of nuclear fission or fusion reactions. In its broadest sense, the term refers both to the uncontrolled release of energy, as in fission or fusion weapons, and to the controlled release of energy, as in a nuclear power plant. Most commonly, however, the expression nuclear power is reserved for the latter of these two processes.

The world's first exposure to nuclear power came when two fission (atomic) bombs were exploded over Hiroshima and Nagasaki, Japan, in August 1945. These actions are said to have brought World War II to a conclusion. After the war, a number of scientists and laypersons looked for some potential peacetime use for this horribly powerful new form of energy. They hoped that the power of nuclear energy could be harnessed to perform work, but those hopes have been realized only to a modest degree. Some serious problems associated with the use of nuclear power have never been satisfactorily solved. As a result, after three decades of progress in the development of controlled nuclear power, interest in this energy source has leveled off and, in many nations, declined.

Words to Know

Cladding: A material that covers the fuel elements in a nuclear reactor in order to prevent the loss of heat and radioactive materials from the fuel.

Coolant: Any material used in a nuclear power plant to transfer the heat produced in the reactor core to another unit in which electricity is generated.

Containment: Any system developed for preventing the release of radioactive materials from a nuclear power plant to the outside world.

Generator: A device for converting kinetic energy (the energy of movement) into electrical energy.

Neutron: A subatomic particle that carries no electrical charge.

Nuclear fission: A reaction in which a larger atomic nucleus breaks apart into two roughly equal, smaller nuclei.

Nuclear fusion: A reaction in which two small nuclei combine with each other to form one larger nucleus.

Nuclear pile: The name given to the earliest form of a nuclear reactor.

Nuclear reactor: Any device for controlling the release of nuclear power so that it can be used for constructive purposes.

Radioactivity: The property possessed by some elements of spontaneously emitting energy in the form of particles or waves by disintegration of their atomic nuclei.

Subatomic particle: Basic unit of matter and energy (proton, neutron, electron, neutrino, and positron) smaller than an atom.

Turbine: A device consisting of a series of baffles (baffles are plates, mounts, or screens that regulate the flow of something—in this case, a liquid) mounted on a wheel around a central shaft. Turbines are used to convert the energy of a moving fluid into the energy of mechanical rotation.

The nuclear power plant

A nuclear power plant is a system in which energy released by fission reactions is captured and used for the generation of electricity. Every

Nuclear power

such plant contains four fundamental elements: the reactor, the coolant system, the electrical power generating unit, and the safety system.

The source of energy in a nuclear reactor is a fission reaction in which neutrons collide with nuclei of uranium-235 or plutonium-239 (the fuel), causing them to split apart. The products of any fission reaction include not only huge amounts of energy, but also waste products, known as fission products, and additional neutrons. A constant and reliable flow of neutrons is ensured in the reactor by means of a moderator, which slows down the speed of neutrons, and control rods, which control the number

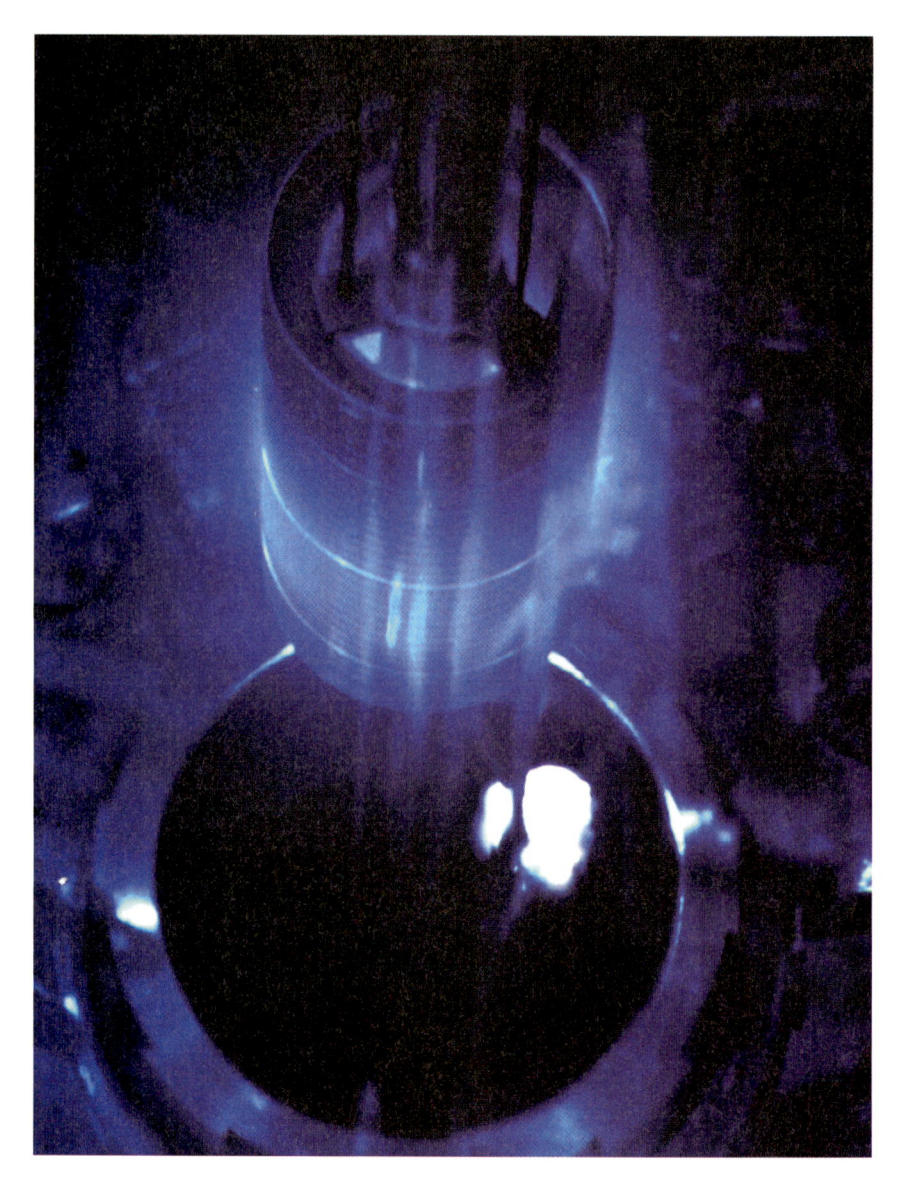

Submerged in water, the fuel element is removed from the reactor at the Oak Ridge National Laboratory. *(Reproduced by permission of Photo Researchers, Inc.)*

of neutrons available in the reactor and, hence, the rate at which fission can occur.

Energy produced in the reactor is carried away by means of a coolant—a fluid such as water, or liquid sodium, or carbon dioxide gas. The fluid absorbs heat from the reactor and then begins to boil itself or to cause water in a secondary system to boil. Steam produced in either of these ways is then piped into the electrical generating unit, where it turns the blades of a turbine. The turbine, in turn, powers a generator that produces electrical energy.

Safety systems. The high cost of constructing a modern nuclear power plant reflects in part the enormous range of safety features needed to protect against various possible mishaps. Some of those features are incorporated into the reactor core itself. For example, all of the fuel in a reactor is sealed in a protective coating made of a zirconium alloy. The protective coating, called a cladding, helps retain heat and radioactivity within the fuel, preventing it from escaping into the power plant itself.

Every nuclear plant is also required to have an elaborate safety system to protect against the most serious potential problem of all: the loss of coolant. If such an accident were to occur, the reactor core might well melt down, releasing radioactive materials to the rest of the plant and, perhaps, to the outside environment. To prevent such an accident from happening, the pipes carrying the coolant are required to be very thick and strong. In addition, backup supplies of the coolant must be available to replace losses in case of a leak.

On another level, the whole plant itself is required to be encased within a dome-shaped containment structure. The containment structure is designed to prevent the release of radioactive materials in case of an accident within the reactor core.

Another safety feature is a system of high-efficiency filters through which all air leaving the building must pass. These filters are designed to trap microscopic particles of radioactive materials that might otherwise be released to the atmosphere. Additional specialized devices and systems have been developed for dealing with other kinds of accidents in various parts of the power plant.

Types of nuclear power plants. Nuclear power plants differ from each other primarily in the methods they use for transferring heat produced in the reactor to the electricity-generating unit. Perhaps the simplest design of all is the boiling water reactor (BWR) plant. In a BWR plant, coolant water surrounding the reactor is allowed to boil and form steam. That steam is then piped directly to turbines, which spin and drive the electrical generator. A very different type of plant is one that was popular in Great Britain for many years—one that used carbon dioxide as a coolant. In this type of plant, carbon dioxide gas passes through the reactor core, absorbs heat produced by fission reactions, and is piped into a secondary system. There the heated carbon dioxide gas gives up its energy to water, which begins to boil and change to steam. That steam is then used to power the turbine and generator.

Safety concerns. In spite of all the systems developed by nuclear engineers, the general public has long had serious concerns about the use of such plants as sources of electrical power. Those concerns vary considerably from nation to nation. In France, for example, more than half of all that country's electrical power now comes from nuclear power plants. By contrast, the initial enthusiasm for nuclear power in the United States in the 1960s and 1970s soon faded, and no new nuclear power plant has been constructed in this country since the mid-1980s. Currently, 104 commercial nuclear power reactors in 31 states generate about 22 percent of the total electricity produced in the country.

One concern about nuclear power plants, of course, is the memory of the world's first exposure to nuclear power: the atomic bomb blasts. Many people fear that a nuclear power plant may go out of control and explode like a nuclear weapon. Most experts insist that such an event is impossible. But a few major disasters continue to remind the public about the worst dangers associated with nuclear power plants. By far the most serious of those disasters was the explosion that occurred at the Chernobyl Nuclear Power Plant near Kiev in Ukraine in 1986.

On April 16 of that year, one of the four power-generating units in the Chernobyl complex exploded, blowing the top off the containment building. Hundreds of thousands of nearby residents were exposed to deadly or damaging levels of radiation and were removed from the area. Radioactive clouds released by the explosion were detected as far away as western Europe. More than a decade later, the remains of the Chernobyl reactor were still far too radioactive for anyone to spend more than a few minutes in the area.

Critics also worry about the amount of radioactivity released by nuclear power plants on a day-to-day basis. This concern is probably of less importance than is the possibility of a major disaster. Studies have shown that nuclear power plants are so well shielded that the amount of radiation to which nearby residents are exposed under normal circumstances is no more than that of a person living many miles away.

In any case, safety concerns in the United States have been serious enough essentially to bring the construction of new plants to a halt. By the end of the twentieth century, licensing procedures were so complex and so expensive that few industries were interested in working their way through the bureaucratic maze to construct new plants.

Nuclear waste management. Perhaps the single most troubling issue for the nuclear power industry is waste management. After a period of time, the fuel rods in a reactor are no longer able to sustain a chain reaction and must be removed. These rods are still highly radioactive, however, and present a serious threat to human life and the environment. Techniques must be developed for the destruction and/or storage of these wastes.

Nuclear wastes can be classified into two general categories: low-level wastes and high-level wastes. The former consist of materials that release a relatively modest level of radiation and/or that will soon decay to a level where they no longer present a threat to humans and the environment. Storing these materials in underground or underwater reservoirs for a few years is usually satisfactory.

The David-Besse Nuclear Power Plant on the shore of Lake Erie in Oak Harbor, Ohio. *(Reproduced by permission of Field Mark Publications.)*

High-level wastes are a different matter. The materials that make up these wastes are intensely radioactive and are likely to remain so for thousands of years. Short-term methods of storage are unsatisfactory because containers would leak and break open long before the wastes were safe.

For more than two decades, the U.S. government has been attempting to develop a plan for the storage of high-level nuclear wastes. At one time, the plan was to bury the wastes in a salt mine near Lyons, Kansas. Objections from residents of the area and other concerned citizens caused that plan to be shelved. More recently, the government decided to construct a huge crypt in the middle of Yucca Mountain in Nevada for the burial of high-level wastes. Again, complaints by residents of Nevada and other citizens have delayed putting that plan into operation. The government insists, however, that Yucca Mountain will eventually become the long-term storage site for the nation's high-level radioactive wastes. Until then, those wastes are in "temporary" storage at nuclear power sites throughout the United States.

History

The first nuclear reactor was built during World War II (1939–45) as part of the Manhattan Project to build an atomic bomb. The reactor was constructed under the direction of Enrico Fermi in a large room beneath the squash courts at the University of Chicago. It was built as the first concrete test of existing theories of nuclear fission.

Until December 2, 1942, when the reactor was first put into operation, scientists had relied entirely on mathematical calculations to determine the effectiveness of nuclear fission as an energy source. It goes without saying that the scientists who constructed the first reactor were taking an extraordinary chance.

That first reactor consisted of alternating layers of uranium and uranium oxide with graphite as a moderator. Cadmium control rods were used to control the concentration of neutrons in the reactor. Since the various parts of the reactor were constructed by piling materials on top of each other, the unit was at first known as an atomic pile. The moment at which Fermi directed the control rods to be withdrawn occurred at 3:45 P.M. on December 2, 1945. That date can legitimately be regarded as the beginning of the age of controlled nuclear power in human history.

Nuclear fusion power

Many scientists believe that the ultimate solution to the world's energy problems may be in the harnessing of nuclear fusion power. A fusion reaction is one in which two small nuclei combine with each other to form one larger nucleus. For example, two hydrogen nuclei may combine with each other to form the nucleus of an atom known as deuterium, or heavy hydrogen.

The world was introduced to the concept of fusion reactions in the 1950s, when the Soviet Union and the United States exploded the first fusion (hydrogen) bombs. The energy released in the explosion of each such bomb was more than 1,000 times greater than the energy released in the explosion of a single fission bomb.

As with fission, scientists and nonscientists alike expressed hope that fusion reactions could someday be harnessed as a source of energy for everyday needs. This line of research has been much less successful, however, than research on fission power plants. In essence, the problem has been to find a way of containing the very high temperatures produced (a few million degrees Celsius) when fusion occurs. Optimistic reports of progress on a fusion power plant appear in the press from time to time, but some authorities now doubt that fusion power will ever be an economic reality.

[See also **Nuclear fission; Nuclear fusion**]

Nuclear weapons

Nuclear weapons are destructive devices that derive their power from nuclear reactions. The term weapon refers to devices such as bombs and warheads designed to deliver explosive power against an enemy. The two types of nuclear reactions used in nuclear weapons are nuclear fission and nuclear fusion. In nuclear fission, large nuclei are broken apart by neutrons, forming smaller nuclei, accompanied by the release of large amounts of energy. In nuclear fusion, small nuclei are combined with each other, again with the release of large amounts of energy.

Fission weapons

The design of a fission weapon is quite simple: all that is needed is an isotope that will undergo nuclear fission. Only three such isotopes exist: uranium-233, uranium-235, and plutonium-239. Fission occurs when the nuclei of any one of these isotopes is struck by a neutron. For example:

neutron + uranium-235 → fission products + energy + more neutrons

The production of neutrons in this reaction means that fission can continue in other uranium-235 nuclei. A reaction of this kind is known as a chain reaction. All that is needed to keep a chain reaction going in

Nuclear weapons

> ### Words to Know
>
> **Fission bomb:** An explosive weapon that uses uranium-235 or plutonium-239 as fuel. Also called an atom bomb.
>
> **Fusion bomb:** An explosive weapon that uses hydrogen isotopes as fuel and an atom bomb as a detonator.
>
> **Isotopes:** Two or more forms of an element that have the same chemical properties but that differ in mass because of differences in the number of neutrons in their nuclei.
>
> **Nuclear fission:** A nuclear reaction in which an atomic nucleus splits into two or more fragments with the release of energy.
>
> **Nuclear fusion:** A nuclear reaction in which two small atomic nuclei combine with each other to form a larger nucleus with the release of energy.
>
> **Radioactivity:** The property possessed by some elements of spontaneously emitting energy in the form of particles or waves by disintegration of their atomic nuclei.

uranium-235 is a block of the isotope of sufficient size. That size is called the critical size for uranium-235.

One of the technical problems in making a fission bomb is producing a block of uranium-235 (or other fissionable material) of exactly the right size—the critical size. If the block is much less than the critical size, neutrons produced during fission escape to the surrounding air. Too few remain to keep a chain reaction going. If the block is larger than critical size, too many neutrons are retained. The chain reaction continues very rapidly and the block of uranium explodes before it can be dropped on an enemy.

The simplest possible design for a fission weapon, then, is to place two pieces of uranium-235 at opposite ends of a weapon casing. Springs are attached to each piece. When the weapon is delivered to the enemy (for example, by dropping a bomb from an airplane), a timing mechanism is triggered. At a given moment, the springs are released, pushing the two chunks of uranium-235 into each other. A piece of critical size is created, fission begins, and in less than a second the weapon explodes.

The only additional detail required is a source of neutrons. Even that factor is not strictly required since neutrons are normally present in the air. However, to be certain that enough neutrons are present to start the fission reaction, a neutron source is also included within the nuclear weapon casing.

Fusion weapons

A fusion weapon obtains the energy it releases from fusion reactions. Those reactions generally involve the combination of four hydrogen atoms to produce one helium atom. Such reactions occur only at very high temperatures, a few million degrees Celsius. The only way to produce temperatures of this magnitude on Earth is with a fission bomb. Thus, a fusion weapon is possible only if a fission bomb is used at its core.

Here is how the fusion bomb is designed: A fission bomb (like the one described in the preceding section) is placed at the middle of the fusion weapon casing. The fission bomb is then surrounded with hydrogen, often in the form of water, since water is two parts hydrogen (H_2O). Even more hydrogen can be packed into the casing, however, if liquid hydrogen is used.

When the weapon is fired, the fission bomb is ignited first. It explodes, releasing huge amounts of energy and briefly raising the temperature inside the casing to a few million degrees Celsius. At this temperature, the hydrogen surrounding the fission bomb begins to fuse, releasing even larger amounts of energy.

The primary advantage that fusion weapons have over fission weapons is their size. Recall that the size of a fission explosion is limited by the critical size of the uranium-235 used in it. A weapon could conceivably consist of two pieces, each less than critical size; or three pieces, each less than critical size; or four pieces, each less than critical size, and so on. But the more pieces used in the weapon, the more difficult the design becomes. One must be certain that the pieces do not come into contact with each other and suddenly exceed critical size.

No such problem exists with a fusion bomb. Once the fission bomb is in place, the casing around it can be filled with ten pounds of hydrogen, 100 pounds of hydrogen, or 100 tons of hydrogen. The only limitation is how large—and heavy—the designer wants the weapon to be.

The power difference between fission and fusion bombs is illustrated by the size of early models of each. The first fission bombs dropped on Japan at the end of World War II were rated as 20 kiloton bombs. The unit kiloton is used to rate the power of a nuclear weapon. It refers to the amount of explosive power produced by a thousand tons of the chemical

Nuclear weapons

explosive TNT. In other words, a 20-kiloton bomb has the explosive power of 20,000 tons of TNT. By comparison, the first fusion bomb ever tested had an explosive power of 5 megatons, or 5 million tons of TNT.

Effects of nuclear weapons

In some respects, the effects produced by nuclear weapons are similar to those produced by conventional chemical explosives. They release heat and generate shock waves. Shock waves are pressure fronts of com-

Computer-enhanced photo of the atom bomb blast over Nagasaki, Japan, on August 8, 1945, that helped bring World War II to a close. (Reproduced by permission of Photo Researchers, Inc.)

pressed air created as hot air expands away from the center of an explosion. They tend to crush objects in their paths. The heat released in a nuclear explosion creates a sphere of burning gas that can range from hundreds of feet to miles in diameter, depending on the power of the bomb. This fireball emits a flash of heat that travels outward from the site of the explosion or ground zero, the area directly under the explosion. The heat from a nuclear blast can set fires and cause serious burns to the flesh of humans and other animals.

Nuclear weapons also produce damage that is not experienced with chemical explosives. Much of the energy released during a weapons blast occurs in the form of X rays, gamma rays, and other forms of radiation that can cause serious harm to plant and animal life. In addition, the isotopes formed during fission and fusion—called fission products—are all radioactive. These fission products are carried many miles away from ground zero and deposited on the ground, on buildings, on plant life, and on animals. As they decay over the weeks, months, and years following a nuclear explosion, the fission products continue to release radiation, causing damage to surrounding organisms.

Nuclear weapons today

Today nuclear weapons are built in many sizes and shapes. They are designed for use against various different types of military and civilian targets. Some weapons are rated at less than 1 kiloton in power, while others have the explosive force of millions of tons of TNT. Small nuclear shells can be fired from cannons. Nuclear warheads mounted on missiles can be launched from land-based silos, ships, submarine, trains, and large-wheeled vehicles. Several warheads can even be fitted into one missile. These MIRVs (or multiple independent reentry vehicles), can release up to a dozen individual nuclear warheads along with decoys far above their targets, making it difficult for the enemy to intercept them.

Even the ability of nuclear weapons to release radioactivity has been exploited to create different types of weapons. Clean bombs are weapons designed to produce as little radioactive fallout as possible. A hydrogen bomb without a uranium jacket would produce relatively little radioactive contamination, for example. A dirty bomb could just as easily be built with materials that contribute to radioactive fallout. Such weapons could also be detonated near Earth's surface to increase the amount of material that could contribute to radioactive fallout. Neutron bombs have been designed to shower battlefields with deadly neutrons that can penetrate buildings and armored vehicles without destroying them. Any *people* exposed to the neutrons, however, would die.

Nuclear weapons

Radioactive Fallout

"The gift that keeps on giving."

That phrase is one way of describing radioactive fallout. Radioactive fallout is material produced by the explosion of a nuclear weapon or by a nuclear reactor accident. This material is blown into the atmosphere and then falls back to Earth over an extended period of time.

Radioactive fallout was an especially serious problem for about 20 years after the first atomic bombs were dropped in 1945. The United States and the former Soviet Union tested hundreds of nuclear weapons in the atmosphere. Each time one of these weapons was tested, huge amounts of radioactive materials were released to the atmosphere. They were then carried around the globe by the atmosphere's prevailing winds. Over long periods of time, they were carried back to Earth's surface or settled to the ground on their own (because of their weight).

More than 60 different kinds of radioactive materials are formed during the explosion of a typical nuclear weapon. Some of these decay and become harmless in a matter of minutes, hours, or days. Other remain radioactive for many years.

An example of a long-lived radioactive material is strontium-90. Strontium-90 loses one-half of its radioactivity every 28 years. It can continue to pose a threat, therefore, for more than a century. The special problem presented by strontium-90 is that it behaves very much like another element—calcium. When it falls to Earth, it is taken up by grass, leaves, and other plant parts. When cows eat grass, they take in strontium-90. The strontium-90 is incorporated into their milk, which is then taken in by humans. Once in the human body, strontium-90 is incorporated into bones and teeth in much the same way that calcium is. Children growing up in the 1960s may still have low levels of strontium-90 in their systems—a "long-lasting gift" from the makers of nuclear weapons.

Nuclear weapons treaties

The United States and Russia signed a Strategic Arms Reduction Treaty (START I) in 1991, which called for the elimination of 9,000 nuclear warheads. Two years later, the two countries signed the START II Treaty, which called for the reduction of an additional 5,000 warheads beyond the number being reduced under START I. Under START II, each country agreed to reduce its total number of strategic nuclear warheads from bombers and missiles by two-thirds by 2003. In 1997, the United States and Russia agreed to delay the elimination deadline until 2007. By that time, each side must have reduced its number of nuclear warheads from 3,000 to 3,500.

Although thousands of nuclear weapons still remain in the hands of many different governments, recent diplomatic trends have helped to lower the number of nuclear weapons in the world. In May 1995, more than 170 members of the United Nations agreed to permanently extend the Nuclear Non-Proliferation Treaty (NPT), which was first signed in 1968. Under terms of the treaty, the five major countries with nuclear weapons—the United States, Britain, France, Russia, and China—agreed to commit themselves to eliminating their arsenals as an ultimate goal and to refusing to give nuclear weapons or technology to any non-nuclear-weapon nation. The other 165 member nations agreed not to acquire nuclear weapons. Israel, which is believed to possess nuclear weapons, did not sign the treaty. Two other nuclear powers, India and Pakistan, refused to renounce nuclear weapons until they can be convinced their nations are safe without them. As of early 2000, a total number of 187 nations had agreed to the NPT. Cuba, India, Israel, and Pakistan were the only nations that had not yet agreed to the treaty.

[*See also* **Nuclear fission; Nuclear fusion**]

Nucleic acid

A nucleic acid is a complex organic compound found in all living organisms. Nucleic acids were discovered in 1869 by the Swiss biochemist Johann Friedrich Miescher (1844–1895). Miescher discovered the presence of an unusual organic compound in the nuclei of cells and gave that compound the name nuclein. The compound was unusual because it contained both nitrogen and phosphorus, in addition to carbon, hydrogen, and oxygen. Nuclein was one of the first organic compounds to have been discovered that contained this combination of elements. Although later research showed that various forms of nuclein occurred in other parts of the cell, the name remained in the modified form by which it is known today: nucleic acid.

Structure of nucleic acids

Nucleic acids are polymers, very large molecules that consist of much smaller units repeated many times over and over again. The small

Nucleic acid

> ## Words to Know
>
> **Amino acid:** One of about two dozen chemical compounds from which proteins are made.
>
> **Cytoplasm:** The fluid inside a cell that surrounds the nucleus and other membrane-enclosed compartments.
>
> **Double helix:** The shape taken by DNA molecules in a nucleus.
>
> **Genetic engineering:** The manipulation of the genetic content of an organism for the sake of genetic analysis or to produce or improve a product.
>
> **Monomer:** A small molecule that can be combined with itself many times over to make a large molecule, the polymer.
>
> **Nitrogen base:** A component of the nucleotides from which nucleic acids are made. It consists of a ring containing carbon, nitrogen, oxygen, and hydrogen.
>
> **Nucleotide:** The basic unit of a nucleic acid. It consists of a simple sugar, a phosphate group, and a nitrogen-containing base.
>
> **Nucleus:** A compartment in the cell that is enclosed by a membrane and that contains its genetic information.
>
> **Phosphate group:** A grouping of one phosphorus atom and four oxygen atoms that occurs in a nucleotide.
>
> **Protein:** A complex chemical compound that consists of many amino acids attached to each other that are essential to the structure and functioning of all living cells.
>
> **Ribosomes:** Small structures in cells where proteins are produced.

units of which polymers are made are known as monomers. In the case of nucleic acid, the monomers are called nucleotides.

The exact structures of nucleotides and nucleic acids are extraordinarily complex. All nucleotides consist of three components: a simple sugar, a phosphate group, and a nitrogen base. A simple sugar is an organic molecule containing only carbon, hydrogen, and oxygen. Perhaps the best-known of all simple sugars is glucose, the sugar that occurs in the blood of mammals and that, when digested, provides energy for their movement. A phosphate group is simply a phosphorus atom to which four oxygen atoms are attached. And a nitrogen base is a simple organic compound that contains nitrogen in addition to carbon, oxygen, and hydrogen.

Kinds of nucleic acids

The term nucleic acid refers to a whole class of compounds that includes dozens of different examples. The phosphate (P) group in all nucleic acids is exactly alike. However, two different kinds of sugars are found in nucleic acids. One kind of sugar is called deoxyribose. The other kind is called ribose. The difference between the two compounds is that deoxyribose contains one oxygen less (deoxy means "without oxygen") than does ribose. Nucleic acids that contain the sugar deoxyribose are called deoxyribonucleic acid, or DNA; those that contain ribose are called ribonucleic acid, or RNA.

Nucleic acids also contain five different kinds of nitrogen bases. The names of those bases and the abbreviations used for them are adenine (A), cytosine (C), guanine (G), thymine (T), and uracil (U). Deoxyribonucleic acids all contain the first four of these nitrogen bases: A, C, G, and T. Ribonucleic acids all contain the first three (A, C, G) and uracil, but not thymine.

DNA and RNA molecules differ from each other, therefore, with regard to the sugar they contain and with regard to the nitrogen bases they contain. They differ in two other important ways: their physical structure and the role they play in living organisms.

Deoxyribonucleic acids (DNA). A single molecule of DNA consists of two very long strands of nucleotides, similar to the structure of all nucleic acids. The two strands are lined up so that the nitrogen bases extending from the sugar-phosphate backbone face each other. Finally, the two strands are twisted around each other, like a pair of coiled telephone cords wrapped around each other. The twisted molecule is known as a double helix.

The function of DNA. One of the greatest discoveries of modern biology occurred in 1953 when the American biologist James Watson (1928–) and the English chemist Francis Crick (1916–) uncovered the role of DNA in living organisms. DNA, Watson and Crick announced, is the "genetic material," the chemical substance in all living cells that passes on genetic characteristics from one generation to the next. How does DNA perform this function?

When a biologist says that genetic characteristics are passed from one generation to the next, one way to understand that statement is to say

Nucleic acid

that offspring know how to produce the same kinds of chemicals they need in their bodies as do their parents. In particular, they know how to produce the most important of all chemicals in living organisms: proteins. Proteins are essential to the function and structure of all living cells.

Watson and Crick said that the way nitrogen bases are lined up in a DNA molecule constitute a kind of "code." The code is not all that different from codes you may use with your friends: A = 1, B = 2, C = 3, and so on. In DNA, however, it takes three nitrogen bases to form a code. For example, the combination CGA means one thing to a cell, the combination GTC another, the combination CCC a third, and so on.

Each possible combination of three nitrogen bases in a DNA molecule stands for one amino acid. Amino acids are the chemical compounds from which proteins are formed. For example, the protein that tells a body to make blue eyes might consist of a thousand amino acids arranged in the sequence A_{15}-A_4-A_{11}-A_8-A_5- and so on. What Watson and Crick said was that every different sequence of nitrogen bases in a DNA molecule stands for a specific sequence of amino acid molecules and, thus, for a specific protein. In the example above, the sequence N_4-N_1-N_2-N_3-N_4-N_3-N_3-N_1-N_4 might conceivably stand for the amino acid sequence A_{15}-A_4-A_{11}-A_8-A_5- which, in turn, might stand for the protein for blue eyes.

When any cell sets about the task of making specific chemicals for which it is responsible, then, it "looks" at the DNA molecules in its nucleus. The code contained in those molecules tells the cell which chemicals to make and how to go about making them.

Ribonucleic acid. So what role do ribonucleic acid (RNA) molecules play in cells? Actually that question is a bit complicated because there are at least three important kinds of RNA: messenger RNA (mRNA); transfer RNA (tRNA); and ribosomal RNA (rRNA). In this discussion, we focus on only the first two kinds of RNA: mRNA and tRNA.

DNA is typically found only in the nuclei of cells. But proteins are not made there. They are made outside the cell in small particles called ribosomes. The primary role of mRNA and tRNA is to read the genetic message stored in DNA molecules in the nucleus, carry that message out of the nucleus and to the ribosomes in the cytoplasm of the cell, and then use that message to make proteins.

The first step in the process takes place in the nucleus of a cell. A DNA molecule in the nucleus is used to create a brand new mRNA molecule that looks almost identical to the DNA molecule. The main difference is that the mRNA molecule is a single long strand, like a long piece

Nucleic acid

of spaghetti. The nitrogen bases on this long strand are a mirror image of the nitrogen bases in the DNA. Thus, they carry exactly the same genetic message as that stored in the DNA molecule.

Once formed, the mRNA molecule passes out of the nucleus and into the cytoplasm, where it attaches itself to a ribosome. The mRNA now simply waits for protein production to begin.

In order for that step to take place, amino acid molecules located throughout the cytoplasm have to be "rounded up" and delivered to the ribosome. There they have to be assembled in exactly the correct order, as determined by the genetic message in the mRNA molecule.

The "carriers" for the amino acid molecules are molecules of transfer RNA (tRNA). Each different tRNA molecule has two distinct ends. One end is designed to seek out and attach itself to some specific amino acid. The other end is designed to seek out and attach itself to some specific sequence of nitrogen bases. Thus, each tRNA molecule circulating in the cell finds the specific amino acid for which it is designed. It attaches itself to that molecule and then transfers the molecule to a ribosome. At the ribosome, the opposite end of the tRNA molecule attaches itself to the mRNA molecule in just the right position. This process is repeated over and over again until every position on the mRNA

A computer-generated model of RNA. *(Reproduced by permission of Photo Researchers, Inc.)*

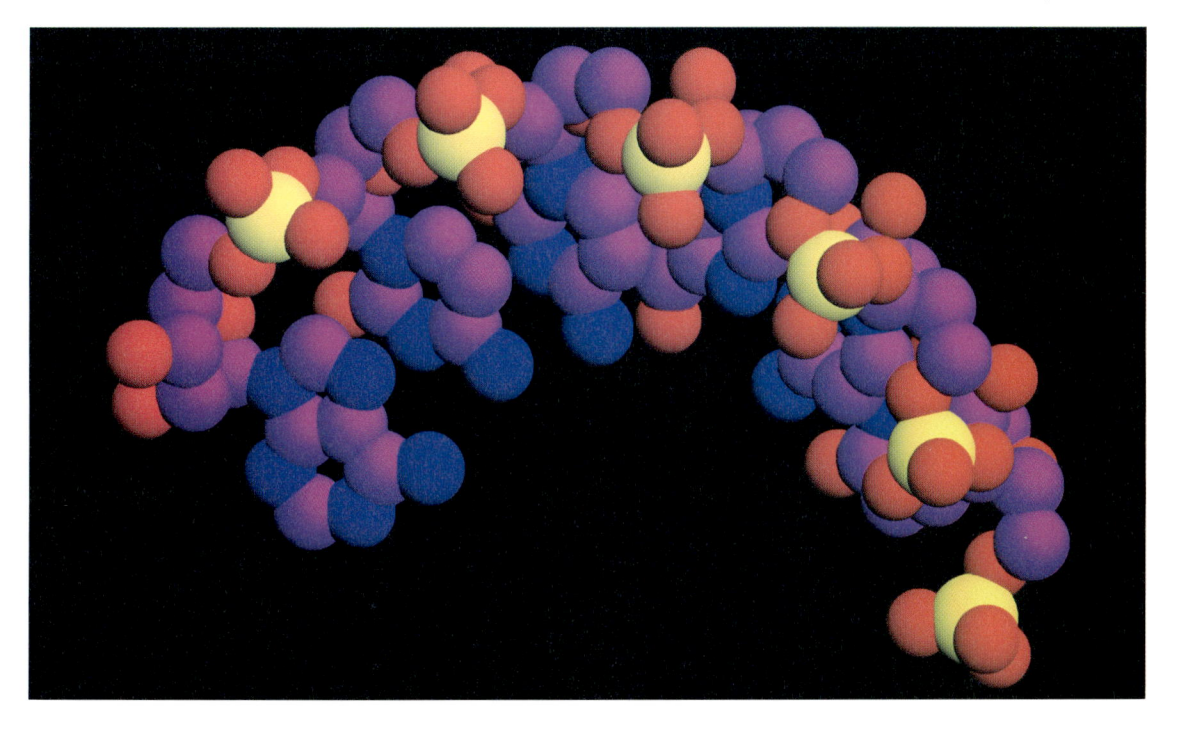

Nucleic acid

molecule holds some specific tRNA molecule. When all tRNA molecules are in place, the amino acids positioned next to each other at the opposite ends of the tRNA molecules join with each other, and a new protein is formed.

Applications

Our understanding of the way in which nucleic acids are constructed and they jobs they do in cells has had profound effects. Today, we can describe very accurately the process by which plant and animal cells learn how to make all the compounds they need to survive, grow, and reproduce. Life, whether it be the life of a plant, a lower animal, or a human, can be expressed in very specific chemical terms.

This understanding has also made possible techniques for altering the way genetic traits are passed from one generation to the next. The process known as genetic engineering, for example, involves making conscious changes in the base sequence in a DNA molecule so that a new set of directions is created and, hence, a new variety of chemicals can be produced by cells.

[*See also* **Chromosome; Enzyme; Genetic engineering; Genetics; Mutation**]

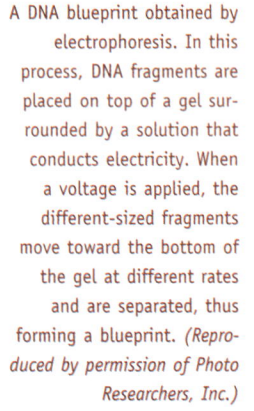

A DNA blueprint obtained by electrophoresis. In this process, DNA fragments are placed on top of a gel surrounded by a solution that conducts electricity. When a voltage is applied, the different-sized fragments move toward the bottom of the gel at different rates and are separated, thus forming a blueprint. *(Reproduced by permission of Photo Researchers, Inc.)*

Number theory

Number theory is the study of natural numbers. Natural numbers are the counting numbers that we use in everyday life: 1, 2, 3, 4, 5, and so on. Zero (0) is often considered to be a natural number as well.

Number theory grew out of various scholars' fascination with numbers. An example of an early problem in number theory was the nature of prime numbers. A prime number is one that can be divided exactly only by itself and 1. Thus 2 is a prime number because it can be divided only by itself (2) and by 1. By comparison, 4 is not a prime number. It can be divided by some number other than itself (that number is 2) and 1. A number that is not prime, like 4, is called a composite number.

The Greek mathematician Euclid (c. 325–270 B.C.) raised a number of questions about the nature of prime numbers as early as the third century B.C. Primes are of interest to mathematicians, for one reason: because they occur in no predictable sequence. The first 20 primes, for example, are 2, 3, 5, 7, 11, 13, 17, 19, 23, 29, 31, 37, 41, 43, 47, 53, 59, 61, 67, and 71. Knowing this sequence, would you be able to predict the next prime number? (It is 73.) Or if you knew that the sequence of primes farther on is 853, 857, 859, 863, and 877, could you predict the next prime? (It is 883.)

Questions like this one have intrigued mathematicians for over 2,000 years. This interest is not based on any practical application the answers may have. They fascinate mathematicians simply because they are engrossing puzzles.

Famous theorems and problems

Studies in number theory over the centuries have produced interesting insights into the properties of natural numbers and ongoing puzzles about such numbers. As just one example of the former, consider Fermat's theorem, a discovery made by French mathematician Pierre de Fermat (1601–1665). Fermat found a quick and easy way to find out if a particular number is a prime or composite number. According to Fermat's theorem, one can determine if any number (call that number p) is a prime number by the following method: choose any number (call that number n) and raise that number to p. Then subtract n from that calculation. Finally, divide that answer by p. If the division comes out evenly, with no remainder, then p is a prime number.

Fermat was also responsible for one of the most famous puzzles in mathematics, his last theorem. This theorem concerns equations of the

Number theory

> ## Words to Know
>
> **Composite number:** A number that can be factored into two or more prime numbers in addition to 1 and itself.
>
> **Cryptography:** The study of creating and breaking secret codes.
>
> **Factors:** Two or more numbers that can be multiplied to equal a product.
>
> **Prime number:** Any number that can be divided evenly only by itself and 1.
>
> **Product:** The number produced by multiplying two or more numbers.

general form $x^n + y^n = z^n$. When n is 2, a very familiar equation results: $x^2 + y^2 = z^2$, the Pythagorean equation of right-angled triangles.

The question that had puzzled mathematicians for many years, however, was whether equations in which n is greater than 2 have any solution. That is, are there solutions for equations such as $x^3 + y^3 = z^3$, $x^4 + y^4 = z^4$, and $x^5 + y^5 = z^5$? In the late 1630s, Fermat wrote a brief note in the margin of a book saying that he had found proof that such equations had no solution when n is greater than 2. He never wrote out that proof, however, and for more than three centuries mathematicians tried to confirm his theory.

As it turns out, any proof that Fermat had discovered was almost certainly wrong. In 1994, Princeton University professor Andrew J. Wiles announced that he had found a solution to Fermat's theorem. But flaws were soon discovered in Wiles's proof (which required more than 150 pages of mathematical equations). By late 1994 Wiles thought the flaws had been solved. However, it will take several years before other mathematicians will be able to verify Wiles's work.

Applications

As mentioned above, the charm of number theory for mathematicians has little or nothing to do with its possible applications in everyday life. Still, such applications do appear from time to time. One such application has come about in the field of cryptography—the writing and deciphering of secret messages (or ciphers). In the 1980s, a number of cryptographers almost simultaneously announced that they had found methods of writing ciphers in such a way that they could be sent across public channels while still remaining secrets. Those methods are based on the fact that it is relatively easy to raise a prime number to some exponent but very difficult to find the prime factors of a large number.

For example, it is relatively simple, if somewhat time-consuming, to find 358^{143}. Actually, the problem is not even time-consuming if a computer is used. However, finding the prime factors of a number such as 384,119,982,448,028 is very difficult unless one knows one of the prime factors to begin with. The way public key cryptography works, then, is to attach some large number, such as 384,119,982,448,028, as a "key" to a secret message. The sender and receiver of the secret message must know one of the prime factors of that number that allows them to decipher the message. In theory, any third party could also decipher the message provided that they could figure out the prime factors of the key. That calculation is theoretically possible although, in practice, it takes thousands or millions of calculations and a number of years, even with the most powerful computers now known.

Numeration systems

Numeration systems are methods for representing quantities. As a simple example, suppose you have a basket of oranges. You might want to keep track of the number of oranges in the basket. Or you might want to sell the oranges to someone else. Or you might simply want to give the basket a numerical code that could be used to tell when and where the oranges came from. In order to perform any of these simple mathematical operations, you would have to begin with some kind of numeration system.

Why numeration systems exist

This example illustrates the three primary reasons that numeration systems exist. First, it is often necessary to tell the number of items contained in a collection or set of those items. To do that, you have to have some method for counting the items. The total number of items is represented by a number known as a cardinal number. If the basket mentioned above contained 30 oranges, then 30 would be a cardinal number since it tells how many of an item there are.

Numbers can also be used to express the rank or sequence or order of items. For example, the individual oranges in the basket could be numbered according to the sequence in which they were picked. Orange #1

Numeration systems

would be the first orange picked; orange #2, the second picked; orange #3, the third picked; and so on. Numbers used in this way are known as ordinal numbers.

Finally, numbers can be used for purposes of identification. Some method must be devised to keep checking and savings accounts, credit card accounts, drivers' licenses, and other kinds of records for different people separated from each other. Conceivably, one could give a name to such records (John T. Jones's checking account at Old Kent Bank), but the number of options using words is insufficient to make such a system work. The use of numbers (account #338-4498-1949) makes it possible to create an unlimited number of separate and individualized records.

History

No one knows exactly when the first numeration system was invented. A notched baboon bone dating back 35,000 years was found in Africa and was apparently used for counting. In the 1930s, a wolf bone was found in Czechoslovakia with 57 notches in several patterns of regular intervals. The bone was dated as being 30,000 years old and is assumed to be a hunter's record of his kills.

The earliest recorded numbering systems go back at least to 3000 B.C., when Sumerians in Mesopotamia were using a numbering system for recording business transactions. People in Egypt and India were using numbering systems at about the same time. The decimal or base-10 numbering system goes back to around 1800 B.C., and decimal systems were common in European and Indian cultures from at least 1000 B.C.

One of the most important inventions in western culture was the development of the Hindu-Arabic notation system (1, 2, 3, . . . 9). That system eventually became the international standard for numeration. The Hindu-Arabic system had been around for at least 2,000 years before the Europeans heard about it, and it included many important innovations. One of these was the placeholding concept of zero. Although the concept of zero as a placeholder had appeared in many cultures in different forms, the first actual written zero as we know it today appeared in India in A.D. 876. The Hindu-Arabic system was brought into Europe in the tenth century with Gerbert of Aurillac (c. 945–1003), a French scholar who studied at Muslim schools in Spain before being named pope (Sylvester II). The system slowly and steadily replaced the numeration system based on Roman numerals (I, II, III, IV, etc.) in Europe, especially in business transactions and mathematics. By the sixteenth century, Europe had largely adopted the far simpler and more economical Hindu-Arabic system of notation, although Roman numerals were still used at times and are even used today.

Numeration systems continue to be invented to this day, especially when companies develop systems of serial numbers to identify new products. The binary (base-2), octal (base-8), and hexadecimal (base-16) numbering systems used in computers were developed in the late 1950s for processing electronic signals in computers.

The bases of numeration systems

Every numeration system is founded on some number as its base. The base of a system can be thought of as the highest number to which one can count without repeating any previous number. In the decimal system used in most parts of the world today, the base is 10. Counting in the decimal system involves the use of ten different digits: 0, 1, 2, 3, 4, 5, 6, 7, 8, and 9. To count beyond 9, one uses the same digits over again—but in different combinations: a 1 with a 0, a 1 with a 1, a 1 with a 2, and so on.

The base chosen for a numeration system often reflects actual methods of counting used by humans. For example, the decimal system may have developed because most humans have ten fingers. An easy way to create numbers, then, is to count off one's ten fingers, one at a time.

Place value

Most numeration systems make use of a concept known as place value. That term means that the numerical value of a digit depends on its location in a number. For example, the number one hundred eleven consists of three 1s: 111. Yet each of the 1s in the number has a different meaning because of its location in the number. The first 1, *1*11, means 100 because it stands in the third position from the right in the number, the hundreds place. (Note that position placement from the right is based on the decimal as a starting point.) The second 1, 1*1*1, means ten because it stands in the second position from the right, the tens place. The third 1, 11*1*, means one because it stands in the first position from the right, the units place.

One way to think of the place value of a digit is as an exponent (or power) of the base. Starting from the right of the number, each digit has a value one exponent larger. The digit farthest to the right, then, has its value multiplied by 10^0 (or 1). The digit next to it on the left has its value multiplied by 10^1 (or 10). The digit next on the left has its value multiplied by 10^2 (or 100). And so forth.

The Roman numeration system is an example of a system without place value. The number III in the Roman system stands for three. Each

Numeration systems

of the 1s has exactly the same value (one), no matter where it occurs in the number. One disadvantage of the Roman system is the much greater difficulty of performing mathematical operations, such as addition, subtraction, multiplication, and division.

Examples of nondecimal numeration systems

Throughout history, numeration systems with many bases have been used. Besides the base 10-system with which we are most familiar, the two most common are those with base 2 and base 60.

Base 2. The base 2- (or binary) numeration system makes use of only two digits: 0 and 1. Counting in this system proceeds as follows: 0; 1; 10; 11; 100; 101; 110; etc. In order to understand the decimal value of these numbers, think of the base 2-system in terms of exponents of base 2. The value of any number in the binary system depends on its place, as shown below:

$$2^3 \ (=8)$$
$$2^2 \ (=4)$$
$$2^1 \ (=2)$$
$$2^0 \ (=1)$$

The value of a number in the binary system can be determined in the same way as in the decimal system.

Anyone who has been brought up with the decimal system might wonder what the point of using the binary system is. At first glance, it seems extremely complicated. One major application of the binary system is in electrical and electronic systems in which a switch can be turned on or off. When you press a button on a handheld calculator, for example, you send an electric current through chips in the calculator. The current turns some switches on and some switches off. If an on position is represented by the number 1 and an off position by the number 0, calculations can be performed in the binary system.

Base 60. How the base-60 numeration system was developed is unknown. But we do know that the system has been widely used throughout human history. It first appeared in the Sumerian civilization in Mesopotamia in about 3000 B.C. Remnants of the system remain today. For example, we use it in telling time. Each hour is divided into 60 minutes and, in turn, each minute into 60 seconds. In counting time, we do not count from 1 to 10 and start over again, but from 1 to 60 before starting over. Navigational systems also use a base-60 system. Each degree of arc on Earth's surface (longitude and latitude) is divided into 60 minutes of arc. Each minute, in turn, is divided into 60 seconds of arc.

Nutrition

The term nutrition refers to the sum total of all the processes by which an organism takes in and makes use of the foods it needs to survive, grow, move, and develop. The word nutrition is also used to refer to the study of the substances an organism needs in order to survive. Those substances are known as nutrients.

Some organisms, such as plants, require nothing other than a supply of light, water, and simple chemicals in order to thrive. Such organisms are known as autotrophs, or self nourishers. Autotrophs build all the molecules they need and capture energy in the process. A few nonplant autotrophic organisms live in the deep oceans near hydrothermal vents (cracks in the ocean floor caused by volcanic activity). These organisms are able to build their own nutrients without using sunlight from sulfur compounds found around the vents.

While green plants get the energy they need directly from sunlight, animals must get the energy they need for life functions from plants.

Nutrients

The major classes of nutrients are carbohydrates, proteins, lipids (or fats), vitamins, and minerals. Animals also need other substances, such as water, fiber, and oxygen, in order to survive. But these substances are not usually regarded as nutrients.

Proteins. Proteins are large molecules built from different combinations of simpler compounds known as amino acids. Human proteins consist of 20 different amino acids. Of these 20 amino acids, the human body is able to manufacture 12 from the foods we eat. The body is unable, however, to make the remaining 8 amino acids it needs for protein production. These 8 amino acids are said to be essential because it is essential that they be included in the human diet.

Proteins that contain all of the essential amino acids are said to be complete proteins. Good sources of complete proteins include fish, meat, poultry, eggs, milk, and cheese. Proteins lacking one or more essential amino acids are incomplete proteins. Peas, beans, lentils, nuts, and cereal grains are sources of incomplete proteins. Anyone whose diet consists primarily

Nutrition

> **Words to Know**
>
> **Amino acid:** A chemical compound used in the construction of proteins.
>
> **Autotroph:** An organism that can build all the food and produce all the energy it needs with its own resources.
>
> **Carbohydrate:** A chemical compound, such as sugar or starch, used by animals as a source of energy.
>
> **Complete protein:** A protein that contains all essential amino acids.
>
> **Edema:** An abnormal collection of fluids in body tissues.
>
> **Essential amino acids:** Amino acids that cannot be produced by an animal, such as a human, and that must, therefore, be obtained from that animal's regular diet.
>
> **Food pyramid:** A diagram developed by the U.S. Department of Agriculture that illustrates the relative amounts of various nutrients needed for normal human growth and development.
>
> **Glycogen:** A chemical compound in which unused carbohydrates are stored in an animal's body.
>
> **Incomplete protein:** A protein that lacks one or more essential amino acids.

of corn and corn products would be at risk for developing health problems because corn lacks two essential amino acids: lysine and tryptophan.

The function of proteins is to promote normal growth, repair damaged tissue, make enzymes, and contribute to the body's immune system.

Carbohydrates. The carbohydrates include sugar and starchy foods, such as those found in cereal grains, potatoes, rice, and fruits. Their primary function in the body is to supply energy. When a person takes in more carbohydrates than his or her body can use, the excess is converted to a compound known as glycogen. Glycogen is stored in liver and muscle tissue and can be used as a source of energy by the body at future times.

Lipids. The term lipid refers to both fats and oils. Lipids serve a number of functions in the human body. Like carbohydrates, they are used to supply energy. In fact, a gram of lipid produces about three times as much energy as a gram of carbohydrate when it is metabolized (burned). The

Nutrition

Indigestible fiber: Fiber that has no nutritional value, but that aids in the normal functioning of the digestive system.

Lipid: A chemical compound used as a source of energy, to provide insulation, and to protect organs in an animal body; a fat or oil

Micronutrient: A nutrient needed in only small amounts by an organism.

Mineral: An inorganic substance found in nature.

Night blindness: Inability to see at night due to a vitamin A deficiency.

Nutrient: A substance needed by an organism in order for it to survive, grow, and develop.

Nutrient deficiency disease: A disease that develops when an organism receives less of a nutrient than it needs to remain healthy.

Protein: A complex chemical compound that consists of many amino acids attached to each other that are essential to the structure and functioning of all living cells.

Vitamin: A complex organic compound found naturally in plants and animals that the body needs in small amounts for normal growth and activity.

release of energy from lipids takes place much more slowly than it does from carbohydrates, however.

Lipids also protect the body's organs from shock and damage and provide insulation for the body.

Vitamins and minerals. Vitamins and minerals are substances needed by the body in only very small amounts. They are also substances that the body cannot produce itself. Thus, they must be included in a person's diet on a regular basis. Vitamins and minerals are sometimes known as micronutrients because they are needed in such small quantities.

An example of a vitamin is the compound known as vitamin A. Vitamin A is required in order for a person to be able to see well at night. An absence of the vitamin can result in a condition known as nightblindness as well as in dryness of the skin. Vitamin A occurs naturally in foods such as green and yellow vegetables, eggs, fruits, and liver.

Nutrition

An example of a mineral is calcium, an element needed to build strong bones and teeth. Calcium is also involved in the normal function of nerve and muscle activity. Good sources of calcium include milk and eggs.

The food pyramid

The food pyramid is a diagram developed by the U.S. Department of Agriculture (USDA) to illustrate the components needed in a healthy diet. The bottom level of the pyramid contains the cereal foods, such as breads, pastas, and rice. This group of foods consists primarily of carbohydrates and is, therefore, a major source of energy. The USDA recommends 6 to 11 servings per day from this group. A serving consists of 30 to 60 grams (1 to 2 ounces) of the food. The exact number of servings depends on the age, gender, weight, and degree of activity for any given person.

The second level of the food pyramid consists of fruits and vegetables. These foods are especially important in supplying vitamins and minerals. A second benefit derived from this group comes from indigestible

Food Guide Pyramid
A Guide to Daily Food Choices

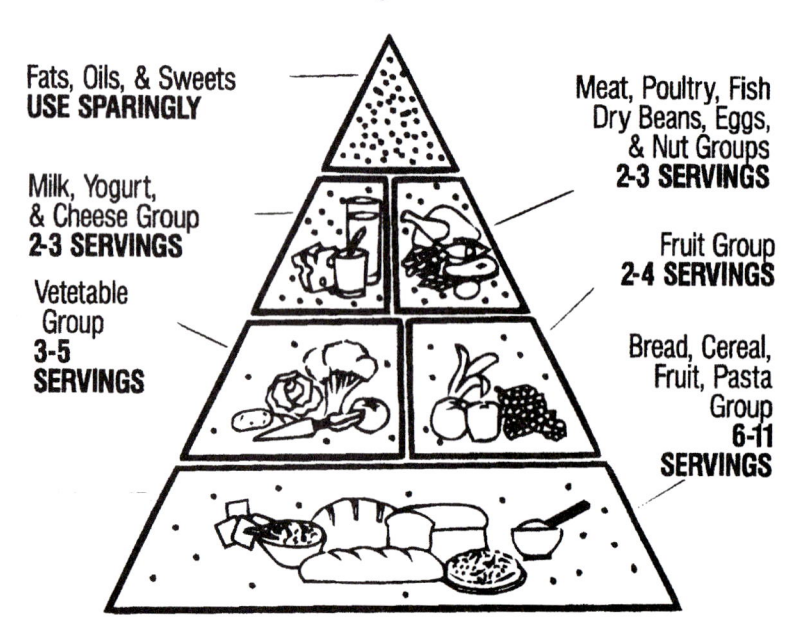

- Fats, Oils, & Sweets **USE SPARINGLY**
- Milk, Yogurt, & Cheese Group **2-3 SERVINGS**
- Vetetable Group **3-5 SERVINGS**
- Meat, Poultry, Fish Dry Beans, Eggs, & Nut Groups **2-3 SERVINGS**
- Fruit Group **2-4 SERVINGS**
- Bread, Cereal, Fruit, Pasta Group **6-11 SERVINGS**

SOURCE: U.S. Department of Agriculture
U.S. Department of Health and Human Services

The food pyramid developed by the U.S. Department of Agriculture. *(Reproduced by permission of the U.S. Department of Agriculture.)*

fiber. Indigestible fiber has been shown to improve the functioning and health of the large intestine. Five to nine servings a day are suggested from this group.

The third level of the pyramid consists of proteins in the form of meats, eggs, beans, nuts, and milk products. This level is smaller than the first and second levels to emphasize that the percentage of these foods should be smaller in comparison to a person's total food intake.

The tip of the pyramid contains the lipids. The small space allotted to the lipids emphasizes that fats and oils should be consumed in small quantities for optimum health.

Nutrient deficiency diseases

The lack of any nutrient can lead to some kind of disease. For example, people who do not have enough protein in their diets may develop a condition known as kwashiorkor. Kwashiorkor (pronounced kwah-shee-OR-kor) is characterized by apathy (lack of interest), muscular wasting, and edema (collection of water in the body). Both the hair and skin lose their pigmentation, and the skin becomes scaly. Diarrhea and anemia (a blood disorder characterized by tiredness) are common, and permanent blindness may result from the condition. Experts estimate that millions of infants die every year worldwide from kwashiorkor.

Rickets is an example of a vitamin deficiency disorder. Rickets develops when a person does not receive enough vitamin D in his or her diet. As a result, the person's bones do not develop properly. His or her legs become bowed by the weight of the body, and wrists and ankles become thickened. Teeth are also badly affected and may take much longer than normal to mature, if they do so at all. Rickets is common among dwellers in slums, where sunlight is not available. (Sunlight causes the natural formation of vitamin D in the skin.) Rickets is no longer a threat in many nations because milk and infant formulas have vitamin D added to them artificially.

[*See also* **Amino acid; Carbohydrate; Lipids; Malnutrition; Protein; Vitamin**]

Obsession

An obsession is a persistent (continuous) and recurring thought that a person is unable to control. A person suffering from obsessive thoughts often has symptoms of anxiety (uneasiness or dread) or emotional distress. To relieve this anxiety, a person may resort to compulsive behavior.

A compulsion is an irresistible impulse or desire to perform some act over and over. Examples of compulsive behavior are repetitive hand washing or turning a light on and off again and again to be certain it is on or off.

Although performing the specific act relieves the tension of the obsession, the person feels no pleasure from the action. On the contrary, the compulsive behavior combined with the obsession cause a great deal of distress for the person. The main concern of psychiatrists and therapists who treat people with obsessions is the role those obsessions play in a mental illness called obsessive-compulsive disorder.

Obsessive-compulsive disorder

Obsessive-compulsive disorder (OCD) is classified as an anxiety disorder. A person suffering from an obsession may be aware of how irrational or senseless their obsession is. However, that person is overwhelmed by the need to perform some repetitive behavior in order to relieve the anxiety connected with the obsession.

OCD makes normal functioning and social interactions very difficult because it tends to consume more and more of a person's time and energy. For example, a person obsessed with the fear of being dirty might

Obsession

> **Words to Know**
>
> **Compulsive behavior:** Behavior that is driven by irresistible impulses to perform some act over and over.
>
> **Flooding:** Exposing a person with an obsession to his or her fears as a way of helping him or her face and overcome them.
>
> **Obsessive-compulsive disorder:** Mental illness in which a person is driven to compulsive behavior to relieve the anxiety of an obsession.
>
> **Obsessive-compulsive personality disorder:** Mental illness in which a person is overtly preoccupied with minor details to the exclusion of larger goals.

spend three to four hours in the bathroom, washing and rewashing himself or herself. Fortunately, OCD is a rare disorder, affecting less than 5 percent of people suffering psychiatric problems.

Obsessive-compulsive personality disorder

People who are overt perfectionists or are rigidly controlling may be suffering from obsessive-compulsive personality disorder (OCPD). In this disorder, the patient may spend excessive amounts of energy on details and lose perspective about the overall goals of a task or job. Obsessive personalities tend to be rigid and unreasonable about how things must be done. They tend also to be workaholics, preferring work over the pleasures of leisure-time activities.

OCPD does not involve specific obsessions or compulsions. The obsessive behavior arises more from generalized attitudes about perfectionism than from a specific obsessive thought. A person suffering from OCPD may be able to function quite successfully at work, but makes everyone else miserable by demanding the same excessive standards of perfection.

Treatments for obsessive-compulsive illnesses

Therapists first try to make patients suffering from obsessive-compulsive illnesses understand that thoughts cannot be controlled. They then try to have patients face the fears that produce their anxiety and gradually learn to deal with them. This type of therapy is called flooding. Once patients begin to modify or change their behavior, they find that the obsessive thoughts begin to diminish.

Most professionals who treat obsessive-compulsive illnesses feel that a combination of therapy and medication is helpful. Some antidepressants, like Anafranil™ and Prozac™, are prescribed to help ease the condition.

Ocean

Oceans are large bodies of salt water that surround Earth's continents and occupy the basins between them. The four major oceans of the world are the Atlantic, Arctic, Indian, and Pacific. These interconnected oceans are further divided into smaller regions of water called seas, gulfs, and bays.

The combined oceans cover almost 71 percent of Earth's surface, or about 139,400,000 square miles (361,000,000 square kilometers). The average temperature of the world's oceans is 39°F (3.9°C). The average depth is 12,230 feet (3,730 meters).

Waves erode the land upon which they land as well as the ocean floor. *(Reproduced by permission of The Stock Market.)*

Ocean

> **Words to Know**
>
> **Continental margin:** Underwater plains connected to continents, separating them from the deep ocean floor.
>
> **Fracture zone:** Faults in the ocean floor that form at nearly right angles to the ocean's major ridges.
>
> **Guyot:** An extinct, submarine volcano with a flat top.
>
> **Ridge:** Very long underwater mountain ranges created as a by-product of seafloor spreading.
>
> **Rift:** Crevice that runs down the middle of a ridge.
>
> **Seafloor spreading:** Process whereby new oceanic crust is created at ridges.
>
> **Seamount:** Active or inactive submarine volcano.

Origin of ocean water

One scientific theory about the origin of ocean water states that as Earth formed from a cloud of gas and dust more than 4.5 billion years ago, a huge amount of lighter elements (including hydrogen and oxygen) became trapped inside the molten interior of the young planet. During the first one to two billion years after Earth's formation, these elemental gases rose through thousands of miles of molten and melting rock to erupt on the surface through volcanoes and fissures (long narrow cracks).

Within the planet and above the surface, oxygen combined with hydrogen to form water. Enormous quantities of water shrouded the globe as an incredibly dense atmosphere of water vapor. Near the top of the atmosphere, where heat could be lost to outer space, water vapor condensed to liquid and fell back into the water vapor layer below, cooling the layer. This atmospheric cooling process continued until the first raindrops fell to the young Earth's surface and flashed into steam. This was the beginning of a fantastic rainstorm that, with the passage of time, gradually filled the ocean basins.

Cosmic rain. In mid-1997, however, scientists offered a new theory on the how the oceans possibly filled in. The National Aeronautics and Space Administration's Polar satellite, launched in early 1996, discovered that small comets about 40 feet (12 meters) in diameter are bombarding Earth's atmosphere at a rate of about 43,000 a day. These comets break up into icy fragments at heights 600 to 15,000 miles (960 to 24,000 kilometers) above ground. Sunlight then vaporizes these fragments into huge clouds, which condense into rain as they sink lower in the atmosphere.

Scientists calculate that this cosmic rain adds one inch of water to Earth's surface every 10,000 to 20,000 years. This amount of water could have been enough to fill the oceans if these comets have been entering Earth's atmosphere since the planet's beginning 4.5 billion years ago.

Ocean basin

Ocean basins are that part of Earth's surface that extends seaward from the continental margins (underwater plains connected to continents, separating them from the deep ocean floor). Basins range from an average water depth of about 6,500 feet (2,000 meters) down into the deepest trenches. Ocean basins cover about 70 percent of the total ocean area.

The familiar landscapes of continents are mirrored, and generally magnified, by similar features in the ocean basin. The largest underwater mountains, for example, are higher than those on the continents. Underwater plains are flatter and more extensive than those on the continents. All basins contain certain common features that include oceanic ridges, trenches, fracture zones, abyssal plains, and volcanic cones.

Oceanic ridges. Enormous mountain ranges, or oceanic ridges, cover the ocean floor. The Mid-Atlantic Ridge, for example, begins at the tip of Greenland, runs down the center of the Atlantic Ocean between the Americas on the west and Africa on the east, and ends at the southern tip of the African continent. At that point, it stretches around the eastern edge of Africa, where it becomes the Mid-Indian Ridge. That ridge continues eastward, making connections with other ridges that eventually end along the western coastline of South and Central America. Some scientists say this is a single oceanic ridge that encircles Earth, one that stretches a total of more than 40,000 miles (65,000 kilometers).

In most locations, oceanic ridges are 6,500 feet (2,000 meters) or more below the surface of the oceans. In a few places, however, they actually extend above sea level and form islands. Iceland (in the North Atlantic), the Azores (about 900 miles [about 1,500 kilometers] off the coast of Portugal), and Tristan de Cunha (in the South Atlantic midway between southern Africa and South America) are examples of such islands.

Ocean

Running along the middle of an oceanic ridge, there is often a deep crevice known as a rift, or median valley. This central rift can plunge as far as 6,500 feet (2,000 meters) below the top of the ridge that surrounds it. Scientists believe ocean ridges are formed when molten rock, or magma, escapes from Earth's interior to form the seafloor, a process known as seafloor spreading. Rifts may be the specific parts of the ridges where the magma escapes.

Trenches. Trenches are long, narrow, canyonlike structures, most often found next to a continental margin. They occur much more commonly in the Pacific than in any of the other oceans. The deepest trench on Earth is the Mariana Trench, which runs from the coast of Japan south and then west toward the Philippine Islands—a distance of about 1,580 miles (2,540 kilometers). Its deepest spot is 36,198 feet (11,033 meters) below sea level. The longest trench is located along the coast of Peru and Chile. Its total length is 3,700 miles (5,950 kilometers) and it has a maximum depth of 26,420 feet (8,050 meters). Earthquakes and volcanic activity are commonly associated with trenches.

Fracture zones. Fracture zones are regions where sections of the ocean floor slide past each other, relieving tension created by seafloor spreading at the ocean ridges. Ocean crust in a fracture zone looks like it has

Underwater ridge of the Juan de Fuca plate off the coast of Washington State. These chimneylike structures on the ocean floor are shaped by emissions of sulfides from deep beneath Earth's crust. *(Reproduced by permission of U.S. Geological Survey Photographic Library.)*

been sliced up by a giant knife. The faults in a zone usually cut across ocean ridges, often nearly at right angles to the ridge. A map of the North Atlantic Ocean basin, for example, shows the Mid-Atlantic Ridge traveling from north to south across the middle of the basin, with dozens of fracture zones cutting across the ridge from east to west.

Abyssal plains. Abyssal plains are relatively flat areas of the ocean basin with slopes of less than one foot of elevation difference for each thousand feet of distance. They tend to be found at depths of 13,000 to 16,000 feet (4,000 to 5,000 meters). Oceanographers believe that abyssal plains are so flat because they are covered with sediments (clay, sand, and gravel) that have been washed off the surface of the continents for hundreds of thousands of years. On the abyssal plains, these layers of sediment have now covered up any irregularities that may exist in the rock of the ocean floor beneath them.

Abyssal plains found in the Atlantic and Indian Oceans tend to be more extensive than those in the Pacific Ocean. One reason for this phenomenon is that the majority of the world's largest rivers empty into either the Atlantic or the Indian Oceans, providing both ocean basins with an endless supply of the sediments from which abyssal plains are made.

Volcanic cones. Ocean basins are alive with volcanic activity. Magma flows upward from the mantle to the ocean bottom not only through rifts, but also through numerous volcanoes and other openings in the ocean floor. Seamounts are submarine volcanoes and can be either active or extinct. Guyots are extinct volcanoes that were once above sea level but have since receded below the surface. As they receded, wave or current action eroded the top of the volcano to a flat surface.

Seamounts and guyots typically rise about 0.6 mile (1 kilometer) above the ocean floor. One of the largest known seamounts is Great Meteor Seamount in the northeastern part of the Atlantic Ocean. It extends to a height of more than 1,300 feet (4,000 meters) above the ocean floor.

[*See also* **Coast and beach; Continental margin; Currents, ocean; Oceanography; Ocean zones; Plate tectonics; Tides; Volcano**]

Oceanography

Oceanography is the scientific study of the oceans, which cover more than 71 percent of Earth's surface. It is divided into four major areas of research: physical, chemical, biological, and geological.

Oceanography

Physical oceanography is the study of basic activities of the oceans such as currents, tides, boundaries, and even evaporation. Chemical oceanography is the study of the chemical parts of the sea and the presence and concentration of chemical elements such as zinc, copper, and nitrogen in the water. These two fields are the main focus of oceanography today. Current areas of research include oceanic circulation—especially ocean currents and their role in weather-related events—and changes in sea level and climate. Also, as the population of the planet continues to increase, oceanographers have begun to conduct research on using the oceans' resources to produce food.

Oceanographers emerge from Alvin, a submersible vessel used for underwater study. (Reproduced by permission of the U.S. Geological Survey Photographic Library, Denver, Colorado.)

Oceanography

Biological oceanography is the study of all life in the sea, including plants, animals, and other living organisms. Since the oceans provide humans with vital food, biological oceanographers look for ways to increase these yields to meet growing populations.

Geological oceanography is the study of the geological structure and mineral content of the ocean floor. This includes mineral resource extraction (removal), seafloor mapping, and plate tectonics activities that offer clues to the origin of Earth. (Earth's crust is made up of large plates that fit together loosely. The study of these plates and their movement, especially their ability to cause earthquakes, is called plate tectonics.)

Ocean research vessels

The era of modern oceanography was opened with the four-year ocean exploration expedition of the H.M.S. *Challenger,* beginning in 1872. The *Challenger* was the first vessel used to systematically record information about all the oceans except the Arctic, including their depths, circulations, temperatures, and organic life.

Sophisticated sonar and magnetic technology on subsequent voyages by other vessels have greatly increased scientists' knowledge of the oceans, helping them make such important geological discoveries as

The *Samuel P. Lee,* a 208-foot (62-meter) oceanographic research vessel carrying out geophysical surveys of the Pacific Ocean and waters around Alaska. *(Reproduced by permission of the U.S. Geological Survey Photographic Library, Denver, Colorado.)*

seafloor spreading (by which new oceanic crust is created) and plate tectonics. All ocean research ships are essentially floating laboratories. Many are operated and financed by the U.S. Navy, often in conjunction with a university or other institution.

Oceanographers of the twenty-first century use satellites to study changes in salt levels, temperature, currents, biological events, and transportation of sediments. As scientists develop new technologies, new doors will open on the study of the world's oceans.

[See also **Ocean**]

Ocean zones

Ocean zones are layers within the oceans that contain distinctive plant and animal life. They are sometimes referred to as ocean layers or environmental zones. The ocean environment is divided into two broad categories, known as realms: the benthic realm (consisting of the seafloor) and the pelagic realm (consisting of the ocean waters). These two realms are then subdivided into separate zones according to the depth of the water.

Water depth versus light penetration

Sunlight obviously cannot penetrate beyond a certain depth in the ocean. Some organisms have, however, evolved to cope with the absence of sunlight at great depths. Plants require sunlight to carry on photosynthesis—the process by which they convert carbon dioxide, water, and other nutrients to simple carbohydrates to produce energy, releasing oxygen as a by-product. Below a depth of about 660 feet (200 meters), not enough sunlight penetrates to allow photosynthesis to occur. The area of the ocean where photosynthesis occurs is known as the euphotic zone (meaning "good light").

From the standpoint of living organisms, the euphotic zone is probably the most important of all oceanic zones. By some estimates, about two-thirds of all the photosynthetic activity that occurs on Earth (on land and in the water) takes place within the euphotic zone.

From 660 to 3,000 feet (200 to 900 meters), only about 1 percent of sunlight penetrates. This layer is known as the dysphotic zone (meaning "bad light"). Below this layer, down to the deepest parts of the ocean, it is perpetual night. This last layer is called the aphotic zone (meaning "without light"). At one time, scientists thought that very little life existed within

Ocean zones

> ### Words to Know
>
> **Benthic:** Pertaining to the ocean floor.
>
> **Consumer:** An organism that consumes other organisms as a food source.
>
> **Chemosynthesis:** The chemical process by which bacteria, by oxidizing hydrogen sulfide, serve as primary producer for a marine community.
>
> **Pelagic:** The water portion of the ocean.
>
> **Photosynthesis:** The process by which green plants produce energy by converting carbon dioxide, water, and other nutrients to simple carbohydrates, releasing oxygen as a by-product.
>
> **Phytoplankton:** Microscopic aquatic plants.
>
> **Producer:** An organism that is capable of utilizing nonliving materials and an external energy source to produce organic molecules (for example, carbohydrates), which are then used as food.
>
> **Zooplankton:** Microscopic aquatic animals.

the aphotic zone. However, they now know that a variety of interesting organisms can be found living on the deepest parts of the ocean floor.

The benthic realm

The benthic realm extends from the shoreline to the deepest parts of the ocean floor. The benthic realm is an especially rich environment for living organisms. Scientists now believe that up to 98 percent of all marine species are found in or near the ocean floor. Some of these are fish or shellfish swimming just above the ocean floor. Most are organisms that burrow in the sand or mud, bore into or are attached to rocks, live in shells, or simply move about on the ocean floor.

In the deeper parts of the ocean floor, below the euphotic zone, no herbivores (plant eaters) can survive. However, the "rain" of dead organic matter from above still supports thriving bottom communities.

The pelagic realm

In the region of the pelagic zone from the surface to 660 feet (200 meters), phytoplankton (algae and microscopic plants) live. They are the

Ocean zones

primary producers of the ocean, the lowest level on the oceanic food web. They use the process of photosynthesis to provide food for themselves and for higher organisms.

On the next level upward in the pelagic food web are the primary consumers, the zooplankton (microscopic animals). They feed on phytoplankton and, in turn, become food for larger animals (secondary consumers) such as sardines, herring, tuna, bonito, and other kinds of fish and swimming mammals. At the top of this food web are the ultimate consumers, the toothed whales.

Humpback whales subsist on some of the smallest ocean creatures: tiny oceanic plankton, or krill. *(Reproduced by permission of the U.S. Fish and Wildlife Service.)*

Ocean zones

In the region from a depth of about 660 to 3,000 feet (200 to 900 meters), a number of organisms survive by spending daylight hours within this region and then rising toward the surface during evening hours. In this way, they can feed off the phytoplankton and zooplankton available near and on the surface of the water while avoiding predators during the day. The most common organisms found in this region are small fish, squid, and simple shellfish. A number of these organisms have evolved some interesting adaptations for living in this twilight world. They often have very large eyes, capable of detecting light only 1 percent as

This hydrothermal vent on the southern Juan de Fuca Ridge is home to a colony of tube worms. *(Reproduced by permission of U.S. Geological Survey Photographic Library.)*

intense as that visible to the human eye. A majority also have light-producing organs that give off a phosphorescence that makes them glow in the dark.

Organisms found below 3,000 feet (900 meters) have also evolved some bizarre adaptations for survival in their lightless environment. In the deeper regions, pressures may exceed 500 times that of atmospheric pressure, or the equivalent of several tons per square inch. Temperatures never get much warmer than about 37°F (3°C). Organisms within these regions generally prey on each other. They have developed special features such as expandable mouths, large and very sharp teeth, and special strategies for hunting or luring prey.

Recent discoveries

In 1977, near the Galapagos Islands in the Pacific Ocean, oceanographers discovered deep sea vents and communities of organisms never seen before. These hydrothermal vents are located in regions where molten rock lies just below the surface of the seafloor, producing underwater hot springs. Volcanic "chimneys" form when the escaping superheated water deposits dissolved minerals and gases upon coming in contact with the cold ocean water. Around these vents are bacteria that obtain energy from the oxidation of hydrogen sulfide escaping from the vents—a process called chemosynthesis. These bacteria (primary producers) are then used as food by tube worms, huge clams, mussels, and other organisms (primary consumers) living around the vents. Since these communities are not photosynthesis-based like all other biological communities, they may provide clues to the nature of early life on Earth.

[See also **Ocean**]

Oil drilling

Oil or petroleum (also known as crude oil) is a fossil fuel found largely in vast underground deposits. Oil and its byproducts (natural gas, gasoline, kerosene, asphalt, and fuel oil, among others) did not have any real economic value until the middle of the nineteenth century when drilling was first used as a method to obtain it. Today, oil is produced on every continent but Antarctica. Despite increasingly sophisticated methods of locating possible deposits and improved removal techniques, oil is still obtained by drilling.

Words to Know

Derrick: The steel tower on a drilling rig or platform that is tall enough to store at least three lengths of 30-foot (9-meter) drill pipe.

Drilling mud: A chemical liquid that cools and lubricates the drill bit and acts as a cap to keep the oil from gushing up.

Fossil fuel: Fuels formed by decaying plants and animals on the ocean floors that were covered by layers of sand and mud. Over millions of years, the layers of sediment created pressure and heat that helped bacteria change the decaying organic material into oil and gas.

Rotary drilling: A drilling system in which the drill bit rotates and cuts into rock.

History

Oil was known in the ancient world and had several uses. Usually found bubbling up to Earth's surface at what are called oil seeps, oil was used primarily for lighting, as a lubricant, for caulking ships (making them watertight), and for jointing masonry (for building). The Chinese knew and used oil as far back as the fourth century B.C.

By the 1850s, crude oil was still obtained by skimming it off the tops of ponds. Since oil from whales was becoming scarce as the giant mammals were hunted almost to extinction, oil producers began to look elsewhere to extract oil. In 1859, while working for the Seneca Oil Company in Titusville, Pennsylvania, Edwin L. Drake and his crew drilled the first modern oil well. They struck oil almost 70 feet (21 meters) down. America's oil boom, and the world's oil industry, was launched.

Oil would be a minor industry for some time, since the only product of crude oil that was thought to be useful was kerosene. The remainder was simply thrown away. Fifty years later, with the invention of the internal-combustion engine and with greater knowledge of the varied applications of petroleum, the oil industry was born in earnest. This market soon became international in scope, and drilling for oil became very serious and sometimes very financially rewarding.

Oil drilling

Drilling for oil

The method Edwin Drake used to drill oil wells is called cable-tool or percussion drilling. A hole is punched into the ground by a heavy cutting tool called a bit that is attached to a cable and pulley system. The cable hangs from the top of a four-legged framework tower called a derrick.

The cable raises and drops the drill bit over and over again, shattering the rock into small pieces or cuttings. Periodically, those cuttings

Deep drilling oil rig at Naval Petroleum Reserve's Elk Hills site near Bakersfield, California. *(Reproduced courtesy of the U.S. Department of Energy.)*

have to be wetted down and bailed out of the hole. By the late 1800s, steam engines had become available for cranking the drill bit up and down and for lowering other tools into the hole.

Although this method is sometimes still used for drilling shallow wells through hard rock, almost all present-day wells are bored by rotary drilling equipment, which works like a corkscrew or carpenter's drill. Rotary drilling originated during the early 1900s in Europe.

Rotary drilling process. In rotary drilling, a large, heavy bit is attached to a length of hollow drill pipe. As the well gets deeper, additional sections of pipe are connected at the top of the hole. The taller the derrick, the longer the sections of drill pipe that can be strung together. Although early derricks were made of wood, modern derricks are constructed of high-strength steel.

The whole length of pipe, or drillstring, is twisted by a rotating turntable that sits on the floor of the derrick. When the drill bit becomes worn, or when a different type of drill bit is needed, the whole drillstring must be pulled out of the hole to change the bit. Each piece of pipe is unscrewed and stacked on the derrick. When the oil-bearing formation is reached, the hole is lined with pipe called casing, and finally the well is completed or made ready for production with cementing material, tubing, and control valves.

Throughout the rotary drilling process, a stream of fluid called drilling mud is continuously forced to the bottom of the hole, through the bit, and back up to the surface. This special mud, which contains clay and chemicals mixed with water, lubricates the bit and keeps it from getting too hot. The drilling mud also carries rock cuttings up out of the hole, clearing the way for the bit and allowing the drilling crew's geologists to study the rock to learn more about the formations underground. The mud also helps prevent cave-ins by shoring up the sides of the hole.

Offshore drilling. Offshore drilling processes and equipment are essentially the same as those on land, except that special types of rigs are used depending on water depth. Jackup rigs, with legs attached to the ocean floor, are used in shallow water with depths to 200 feet (61 meters). In depths up to 4,000 feet (1,220 meters), drilling takes place on semisubmersible rigs that float on air-filled legs and are anchored to the bottom. Drillships with very precise navigational instruments are used in deep water with depths to 8,000 feet (2,440 meters). Once a promising area has been identified, a huge fixed platform is constructed that can support as many as 42 offshore wells, along with living quarters for the drilling crew.

Oil spills

Many advancements have been made in oil-drilling technology. The most advanced rotary cone rock bits presently available can drill about 80 percent faster than bits from the 1920s. At that time, well depths reached about 8,200 feet (2,500 meters). Today's drills can reach down more than 30,000 feet (9,150 meters).

[*See also* **Natural gas; Oil spills; Petroleum**]

Oil spills

Crude oil or petroleum is an important fossil fuel. Fossil fuels were formed millions of years ago, when much of Earth was covered by water containing billions of tiny plants and animals. After these organisms died they accumulated on ocean floors, where, over the ages, sand and mud also drifted down to cover them. As these layers piled up over millions of years, their weight created pressure and heat that changed the decaying organic material into oil and gas.

Since petroleum is often extracted in places that are far away from areas where it is refined and eventually used, it must be transported in

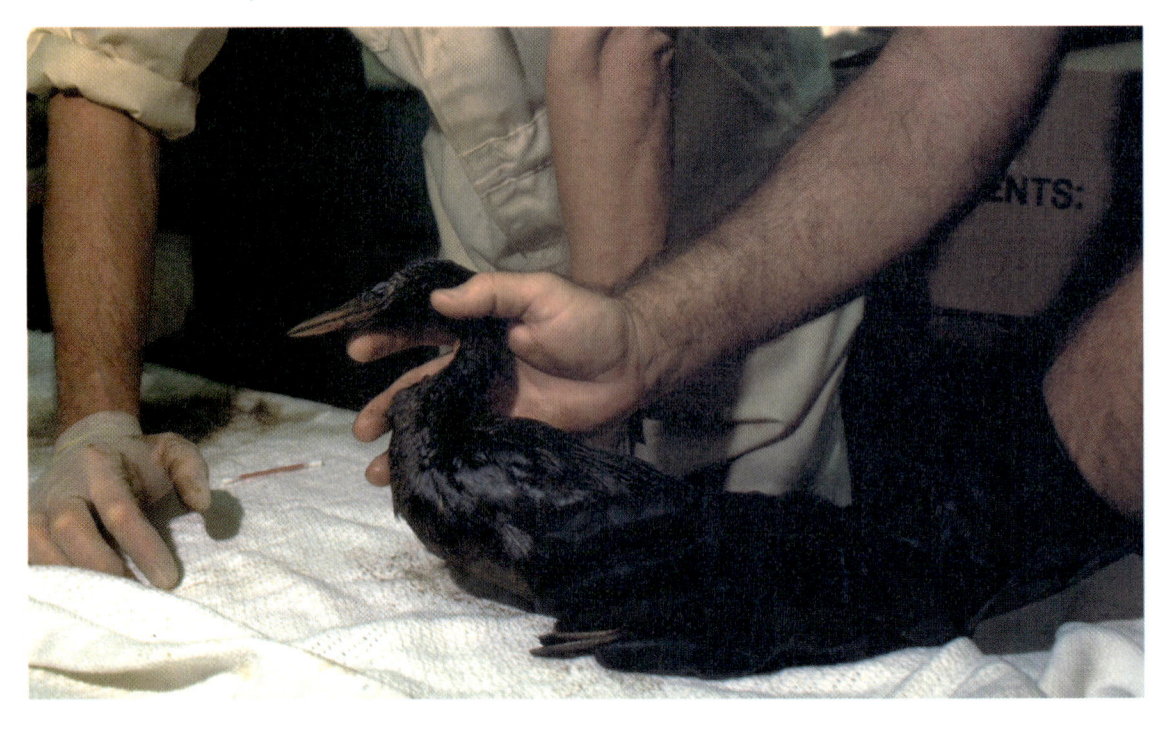

A bird soaked in oil that was spilled into the Persian Gulf in 1991 during the Persian Gulf War. *(Reproduced by permission of Photo Researchers, Inc.)*

Bioremediation

Bioremediation is the use of bacteria and other living microorganisms to help clean up hazardous waste in soil and water. These microorganisms work by decomposing or breaking down toxic chemicals into simpler, less-poisonous compounds. Petroleum can be consumed by microorganisms, transforming it into compounds such as carbon dioxide and water.

Unfortunately, the technique of adding bacteria to a hazardous waste spill in order to increase the rate of decomposition has yielded limited success. If a community of bacteria present in an area where a spill occurs can consume that hazardous waste, the bacteria grow rapidly. Adding further microorganisms does not seem to increase the overall rate of decomposition. More important, areas where spills take place often do not have optimum conditions under which the bacteria can function. Oxygen and certain nutrients such as nitrate and phosphate that the bacteria need are often limited.

However, in the case of the Exxon Valdez oil spill in Alaska in 1989, nitrogen- and phosphorus-containing fertilizer was applied to about 75 miles (120 kilometers) of oiled beach. The fertilizer adhered to the petroleum residues. With these added nutrients, the naturally occurring community of bacteria broke down the oil at a 50 percent faster rate.

large quantities by ocean tankers, inland-water barges, or overland pipelines. Accidental oil spills can occur at any time during the loading, transportation, and unloading of oil. Some oil spills have been spectacular in magnitude and have caused untold environmental damage.

Characteristics of petroleum

Petroleum is a thick, flammable, yellow-to-black colored liquid containing a mixture of organic chemicals, most of which are hydrocarbons (organic compounds containing only hydrogen and carbon atoms). Since petroleum is a natural material, it can be diluted or decomposed by bacteria and other natural agents. The most toxic or poisonous components of petroleum are the volatile components (those compounds that evaporate at low temperatures) and the components that are water-soluble (able to be dissolved in water).

Oil spills

The products refined from petroleum (gasoline, kerosene, asphalt, fuel oil, and petroleum-based chemical), however, are not natural. Because of this, few natural agents like bacteria exist to decompose or break them down.

Oil pollution

Since crude oil floats, oil spilled on the open ocean most severely affects seabirds. The oil coats their bodies, preventing them from fluffing their feathers to keep warm. Animals and plants below the surface of the water are largely unaffected by the oil spill. Naturally present bacteria in the water begin to break down the petroleum, using it as a nutrient. Eventually, the oil slick is broken down into a hard, tarlike substance, which is almost completely harmless to seagoing life of all kind. Wave action then breaks the pieces of tar into progressively smaller and smaller pieces. These small pieces, often invisible to the naked eye, remain floating in the water indefinitely.

When an oil spill occurs near a coastline, however, the damaging effect is much greater. Waves wash the oil on shore where it covers everything—rocks, plants, animals, and sand. If the wave action is high, bacteria present in the water are able to break down the oil fairly rapidly. In areas where wave action is low, such as in a bay or estuary, plants and animals quickly suffocate or are killed by toxic reactions to the oil, which remains for a considerable time. Since many small bays serve as nesting sites for aquatic birds and animals, spills in these areas are especially deadly.

Incidents of oil spills

Accidental oil spills from tankers and offshore rigs are estimated to total about 250 million gallons (950 million liters) per year. In 1967, off the coast of southern England, the *Torrey Canyon* ran aground, spilling about 32 million gallons (120 million liters) of crude oil. In 1978, the *Amoco Cadiz* went aground in the English Channel, spilling over 63 million gallons (240 million liters) of crude oil. In 1979, the offshore exploration rig *IXTOC-I* had an uncontrollable blowout that spilled more than 137 million gallons (520 million liters) of crude oil into the Gulf of Mexico.

The most damaging oil spill ever to occur in North American waters was the Exxon *Valdez* accident of 1989. The 11 million gallons (42 million liters) of spilled oil affected about 1,180 miles (1,900 kilometers) of shoreline of Prince William Sound and its surroundings in Alaska. The coastal habitats of many animals and birds were destroyed. Large numbers of sea mammals and birds were also affected in offshore waters. Of the estimated 5,000 to 10,000 sea otters that lived in Prince William Sound, at least 1,000 were killed by oiling. About 36,000 dead seabirds of various species were collected from beaches and other places. The actual number of killed birds, however, was probably in the range of 100,000 to 300,000. At least 153 bald eagles died from poisoning when they ate the remains of oiled seabirds.

Large quantities of crude oil have also been spilled during warfare. The largest-ever spill of petroleum into a body of water occurred during the Persian Gulf War in 1991–92. Iraqi forces deliberately released an estimated 500 million gallons (1,900 million liters) of petroleum into the Persian Gulf from several tankers and an offshore taker-loading facility. Iraqi forces also sabotaged and set fire to over 500 oil wells in Kuwait. Before the fires were extinguished and the wells capped, an estimated 10 to 30 billion gallons (35 to 115 billion liters) of crude oil had been spilled. Much of the oil burned, releasing toxic petroleum vapors into the atmosphere.

A potential ecological disaster arose in January 2001 when a tanker carrying 234,000 gallons (885,690 liters) of diesel and bunker (a heavy mixed fuel used by tourist boats) spilled the bulk of its load after it ran aground near the Galápagos Islands. The islands are located in the Pacific Ocean about 600 miles (1,000 kilometers) west of the country of Ecuador,

Oil spills

Clean-up of the Exxon *Valdez* oil spill in Prince William Sound, Alaska, using hot and cold water jets. *(Reproduced by permission of Greenpeace Photos.)*

Orbit

of which they are a province. In fact, the islands are Ecuador's main tourist attraction. The islands are an ecosystem populated by species found nowhere else in the world. With their rare species of birds, reptiles, and marine life, the islands were an inspiration for English naturalist Charles Darwin's (1809–1882) theory of evolution.

The tanker hit bottom 550 yards (503 meters) off San Cristóbol, the easternmost island in the Galápagos. Luckily for the islands and their species, strong ocean currents washed almost all of the fuel out to sea, where most of it evaporated. Only two pelicans were found dead on the islands, while dozens were soiled, along with several sea lions and pups and exotic blue-footed booby birds. According to experts, the contamination was minimal. They also felt there would be no long-term damage to the islands from the spill.

[*See also* **Petroleum; Pollution**]

Orbit

An orbit is the path a celestial object follows when moving under the control of another's gravity. This gravitational effect is evident throughout the universe: satellites orbit planets, planets orbit stars, stars orbit the cores of galaxies, and galaxies revolve in clusters.

Without gravity, celestial objects would hurtle off in all directions. Gravity pulls those objects into circular and elliptical (oval-shaped) orbits. Indeed, gravity was responsible for the clumping together of dust and gas shortly after the beginning of the universe, which led to the formation of stars and galaxies.

Kepler's laws and planetary motion

Since ancient times, astronomers have been attempting to understand the patterns in which planets travel throughout the solar system and the forces that propel them. One such astronomer was the German Johannes Kepler (1571–1630). In 1595, he discovered that the planets formed ellipses in space. In 1609, he published his first two laws of planetary motion. The first law states that a planet travels around the Sun on an elliptical path. The second law states that a planet moves faster on its orbit when it is closer to the Sun and slower when it is farther away.

Ten years later, Kepler added a third law of planetary motion. This law makes it possible to calculate a planet's relative distance from the

Sun knowing its period of revolution. Specifically, the law states that the cube of the planet's average distance from the Sun is equal to the square of the time it takes that planet to complete its orbit.

Scientists now know that Kepler's planetary laws also describe the motion of stars, moons, and human-made satellites.

Newton's laws

More than 60 years after Kepler published his third law, English physicist Isaac Newton (1642–1727) developed his three laws of motion and his law of universal gravitation. Newton was the first to apply the notion of gravity to orbiting bodies in space. He explained that gravity was the force that made planets remain in their orbits instead of falling away in a straight line. Planetary motion is the result of movement along a straight line combined with the gravitational pull of the Sun.

Newton discovered three laws of motion, which explained interactions between objects. The first is that a moving body tends to remain in motion and a resting body tends to remain at rest unless acted upon by an outside force. The second states that any change in the acceleration of an object is proportional to, and in the same direction as, the force acting on it. (Proportional means corresponding, or having the same ratio.) In addition, the effects of that force will be inversely proportional (opposite) to the mass of the object; that is, when affected by the same force, a heavier object will move slower than a lighter object. Newton's third law states that for every action there is an equal and opposite reaction.

Newton used these laws to develop the law of universal gravitation. This law states that the gravitational force between any two objects depends on the mass of each object and the distance between them. The greater each object's mass, the stronger the pull, but the greater the distance between them, the weaker the pull. The strength of the gravitational force, in turn, directly affects the speed and shape of an object's orbit. As strength increases, so does the orbital speed and the tightness of the orbit.

Newton also added to Kepler's elliptical orbit theory. Newton found that the orbits of objects going around the Sun could be shaped as circles, ellipses, parabolas, or hyperbolas. As a result of his work, the orbits of the planets and their satellites could be calculated very precisely. Scientists used Newton's laws to predict new astronomical events. Comets and planets were eventually predicted and discovered through Newtonian or celestial mechanics—the scientific study of the influence of gravity on the motions of celestial bodies.

Organic chemistry

Einstein revises Newton's laws

In the early 1900s, German-born American physicist Albert Einstein (1879–1955) presented a revolutionary explanation for how gravity works. Whereas Newton viewed space as flat and time as constant (progressing at a constant rate—not slowing down or speeding up), Einstein described space as curved and time as relative (it can slow down or speed up).

According to Einstein, gravity is actually the curvature of space around the mass of an object. As a lighter object (like a planet) approaches a heavier object (like the Sun) in space, the lighter object follows the lines of curved space, which draws it near the heavier object. To understand this concept, imagine space as a huge stretched sheet. If you were to place a large heavy ball on the sheet, it would cause the sheet to sag. Now imagine a marble rolling toward the ball. Rather than traveling in a straight line, the marble would follow the curves in the sheet caused by the ball's depression.

Einstein's ideas did not prove Newton wrong. Einstein merely showed that Newtonian mechanics work more accurately when gravity is weak. Near stars and black holes (single points of infinite mass and gravity that are the remains of massive stars), where there are powerful gravitational fields, only Einstein's theory holds up. Still, for most practical purposes, Newton's laws continue to describe planetary motions well.

[*See also* **Celestial mechanics; Moon; Satellite; Solar system; Star; Sun**]

Organic chemistry

Organic chemistry is the study of compounds of carbon. The name organic goes back to a much earlier time in history when chemists thought that chemical compounds in living organisms were fundamentally different from those that occur in nonliving things. The belief was that the chemicals that could be extracted from or that were produced by living organisms had a special "vitalism" or "breath of life" given to them by some supernatural being. As such, they presented fundamentally different kinds of problems than did the chemicals found in rocks, minerals, water, air, and other nonliving entities. The chemical compounds associated with living organisms were given the name organic to emphasize their connection with life.

In 1828, German chemist Friedrich Wöhler (1800–1882) proved that this theory of vitalism was untrue. He found a very simple way to con-

Major Organic Families and the Functional Groups They Contain

Family	Functional Group
Alkane	carbon–carbon single bonds only: C–C
Alkene	at least one carbon–carbon double bond: C=C
Alkyne	at least one carbon–carbon triple bond: C≡C
Alcohol	hydroxyl group: C–OH
Ether	carbon–oxygen–carbon bonding: C–O–C
Aldehyde and ketone	carboxyl (C=O) group: C–C=O

Carboxylic acid	carboxylic (C=O) group: C – C=O
	                              | |
	                            OH OH

Ester	ester (C=O) group: C–C=O
	                      | |
	                      O O

Amine	amine (NH_2) group: C–NH_2

vert chemical compounds from living organisms into comparable compounds from nonliving entities.

As a result of Wöhler's research, the definition of organic chemistry changed. The new definition was based on the observation that every compound discovered in living organisms had one property in common: they all contained the element carbon. As a result, the modern definition of organic chemistry—as the study of carbon compounds—was adopted.

Organic and inorganic chemistry

One important point that Wöhler's research showed was that the principles and techniques of chemistry apply equally well to compounds found in living organisms and in nonliving things. Nonetheless, some important differences between organic and inorganic (not organic) compounds exist. These include the following:

1. The number of organic compounds vastly exceeds the number of inorganic compounds. The ratio of carbon-based compounds to non-carbon-based compounds is at least ten to one, with close to 10 million organic compounds known today. The reason for this dramatic difference is a special property of the carbon atom: its ability to join with other carbon atoms in very long chains, in rings, and in other kinds of geometric arrangements. It is not at all unusual for dozens, hundreds, or thousands

Organic chemistry

of carbon atoms to bond to each other within a single compound—a property that no other element exhibits.

2. In general, organic compounds tend to have much lower melting and boiling points than do inorganic compounds.

3. In general, organic compounds are less likely to dissolve in water than are inorganic compounds.

4. Organic compounds are likely to be more flammable but poorer conductors of heat and electricity than are inorganic compounds.

5. Organic reactions tend to take place more slowly and to produce a much more complex set of products than do inorganic reactions.

Functional groups and organic families

The huge number of organic compounds requires that some system be developed for organizing them. The criterion on which those compounds are organized is the presence of various functional groups. A functional group is an arrangement of atoms that is responsible for certain characteristic physical and chemical properties in a compound. For example, one such functional group is the hydroxyl group, consisting of an oxygen atom and hydrogen atom joined to each other. It is represented by the formula —OH.

All organic compounds with the same functional group are said to belong to the same organic family. Any organic compound that contains a hydroxyl group, for instance, is called an alcohol. All alcohols are similar to each other in that: (1) they contain one or more hydroxyl groups, and (2) because of those groups, they have similar physical and chemical properties. For example, alcohols tend to be more soluble in water than other organic compounds because the hydroxyl groups in the alcohol form bonds with water molecules.

The simplest organic compounds are the hydrocarbons, compounds that contain only two elements: carbon and hydrogen. The class of hydrocarbons can be divided into subgroups depending on the way in which carbon and hydrogen atoms are joined to each other. In some hydrocarbons, for example, carbon and hydrogen atoms are joined to each other only by single bonds. A single bond is a chemical bond that consists of a pair of electrons. Such hydrocarbons are known as saturated hydrocarbons.

In other hydrocarbons, carbon and hydrogen atoms are joined to each other by double or triple bonds. A double bond consists of two pairs of electrons, and a triple bond consists of three pairs of electrons. The symbols used for single, double, and triple bonds, respectively, are —, =, and ≡. Hydrocarbons containing double and triple bonds are said to be unsaturated.

Hydrocarbons can also be open-chain or ring compounds. In an open-chain hydrocarbon, the carbon atoms are all arranged in a straight line, like a strand of spaghetti. In a ring hydrocarbon, the carbons are arranged in a continuous loop, such as a square, a pentagon, or a triangle.

Organic farming

Organic farming is the process by which crops are raised using only natural methods to maintain soil fertility and to control pests. The amount of crops produced by conventional farming methods is often larger than that of organic farming. But conventional farming, with its heavy use of manufactured fertilizers and pesticides (agrochemicals), has a greater negative effect on the environment. In comparison, organic farming produces healthy crops while maintaining the quality of the soil and surrounding environment.

Soil fertility

Soil fertility is a measure of the soil's ability to grow crops and plants. Fertility is affected by a soil's tilth and the amount of nutrients it contains. Tilth refers to the physical structure of soil. Good tilth means that soil is loose and not compacted. It holds a great amount of water without becoming soggy and permits air to penetrate to plant roots and soil organisms. It also allows plant roots to grow and penetrate deeper.

The nutrients in soil are directly related to the soil's concentration of organic matter (living or dead plants and animals). Plants require more than 20 nutrients for proper growth. Some of these nutrients are obtained primarily from the soil, especially inorganic compounds of nitrogen, phosphorus, potassium, calcium, magnesium, and sulfur. In natural ecosystems, microorganisms (bacteria and fungi) in the soil break down organic matter, releasing the inorganic nutrients necessary for plant growth.

In conventional farming, soil tilth is destroyed by the use of heavy machinery, which compacts the soil. Very little organic material is added to the soil in conventional farming, decreasing the amount of nutrients that are naturally produced. Instead, inorganic nutrients are added directly to the soil in the form of synthetic fertilizers, which are manufactured from raw materials. These fertilizers are often applied at an excessive rate.

Organic farming

> **Words to Know**
>
> **Fertilizers:** Substances added to agricultural lands to encourage plant growth and higher crop production.
>
> **Organic matter:** Any biomass of plants or animals, living or dead.
>
> **Pesticides:** Substances used to reduce the abundance of pests, any living thing that causes injury or disease to crops.
>
> **Tilth:** The physical structure of soil.

As a result, they pass through the soil to contaminate groundwater and flow along the surface of soil to pollute surrounding bodies of water, threatening native species.

In contrast, organic farmers try to increase soil fertility by increasing the organic matter in the soil. They do so by adding the dung and urine of animals (which contains both organic matter and large concentrations of nutrients), by plowing under growing or recently harvested plants (such as alfalfa or clover), or by adding compost or other partially decomposed plants. These methods rely more heavily on renewable sources of energy and materials rather than on nonrenewable materials and fossil fuels.

Managing pests

In agriculture, pests are any living thing that causes injury or disease to crops. This can occur when insects eat foliage or stored produce, when bacteria or fungi cause plant diseases, or when weeds interfere excessively with the growth of crop plants. In conventional farming, pests are usually managed using various types of pesticides, such as insecticides, herbicides, and fungicides. In the short term, these methods can be effective in reducing the influence of pests on crops. However, the long-term use of these chemicals has been shown to have a severe effect on the environment.

Organic farmers do not use synthetic, manufactured pesticides to manage their pest problems. Rather, they rely on other methods. These include using crop varieties that are resistant to pests and diseases, introducing natural predators of the pests, changing the habitat of the crop area to make it less suitable for the pest, and (when necessary) using a pesticide derived from a natural product.

Organic farming

Animal husbandry

In conventional farming, livestock animals are generally kept together under extremely crowded and foul conditions. Because of this, they are highly susceptible to diseases and infections. To manage this problem, conventional farmers rely on antibiotics, which are given not only when animals are sick but often on a continued basis in the animals' feed. Since the mid-1990s, however, scientists have known that this practice has led to the development of new strains of bacteria that are resistant to the repeated use of antibiotics. These bacteria are not only harmful to the animals but are potentially harmful to the humans who consume the animals.

Organic farmers might also use antibiotics to treat infections in sick animals, but they do not continuously add those chemicals to the animals' feed. In addition, many organic farmers keep their animals in more open and sanitary conditions. Animals that are relatively free from crowding and constant exposure to waste products are more resistant to diseases. Overall, they have less of a need for antibiotics.

Intensively managed agriculture (left) compared with organic farming (right). *(Reproduced by permission of The Gale Group.)*

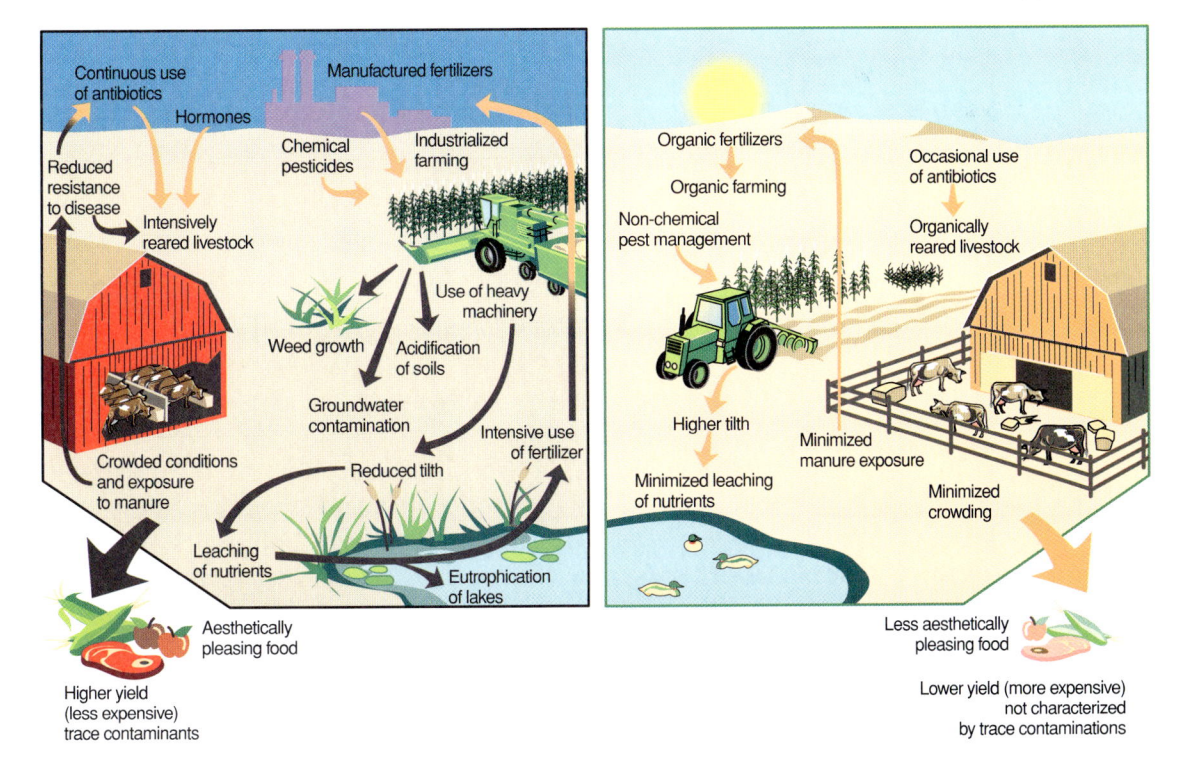

Some conventional farmers raising livestock use synthetic growth hormones, such as bovine growth hormone, to increase the size and productivity of their animals. Inevitably, these hormones remain in trace concentrations, contaminating the animal products that humans consume. Although risk to humans has yet to be scientifically demonstrated, there is controversy about the potential effects. Organic farmers do not use synthetic growth hormones to enhance their livestock.

[*See also* **Agriculture; Agrochemical; Crops; Slash-and-burn agriculture**]

Orthopedics

Orthopedics is the branch of medicine that specializes in diseases of and injuries to bones. French physician Nicholas Andry coined the term "orthopedia" in his 1741 book on the prevention and correction of muscular and skeletal deformities in children. He united the Greek word "orthos," meaning straight, with "pais," meaning child. The term orthopedics has remained in use, though the specialty has broadened beyond the care of children.

Bone is a living and functioning part of the body. A broken bone will generate new growth to repair the fracture and fill in any areas from which bone is removed. Therefore, a bone that is deformed from birth can be manipulated, cut, braced, or otherwise treated to produce a normal form. A broken bone held in alignment will heal with no resulting physical deformity.

History of orthopedics

Humans have had to contend with broken or malformed bones since prehistory. Ancient Egyptian hieroglyphics (system of writing in which pictures or symbols represent words or sounds) depict injured limbs wrapped and braced to heal normally. As wars were waged on a larger scale and weaponry became more efficient and deadly over time, fractures and other bone injuries became more common. Physicians soon developed simple prostheses (pronounced pros-THEE-sees; artificial limbs) to replace limbs that were amputated as the result of a wound. A hand, for example, was replaced with a hook attached to a cup that fit over the wrist.

Early orthopedists (orthopedic physicians) concentrated on the correction of birth defects such as scoliosis (abnormal sideways curvature of the spine) and clubfoot (deformed foot marked by a curled or twisted

> ## Words to Know
>
> **Clubfoot:** Deformed foot marked by a curled or twisted shape.
>
> **Prosthesis:** Artificial device to replace a missing part of the body.
>
> **Scoliosis:** An abnormal sideways curvature of the spine.

shape). Gradually orthopedists included fractures, dislocations, and trauma to the spine and skeleton within their specialty.

For many years, orthopedics was a physical specialty. The orthopedist manipulated bones and joints to restore alignment, and then applied casts or braces to maintain the structure until it healed. Fractures of the hip, among other injuries, were considered untreatable and were ignored. The patient was simply made as comfortable as possible while the fracture healed on its own. Often the healing process was not complete, and the patient was left with a lifelong handicap that made walking or bending difficult.

Modern developments in orthopedics

In the 1930s, a special nail was developed to hold bone fragments together to allow them to heal better. A few years afterward, a metal device was invented to replace the head of a femur (thigh bone) that formed part of the hip joint and that often would not heal after being fractured. Later, a total artificial hip joint was invented. It continues to be revised and improved to allow a patient maximum use and flexibility of the leg.

Current orthopedic specialists continue to apply physical methods to align fractures and restore a disrupted joint. Braces and casts are still used to hold injured bones in place while they heal. Now, however, the physician can take X rays to be certain that the bones are aligned properly for healing to take place. X rays also can be taken during the healing process to make sure that the alignment has not changed and that healing is occurring swiftly.

Orthopedists treat crushed bones (which have little chance of healing on their own) by transplanting bits of bone from other locations in the body to fill splintered areas. The operating room in which an orthopedic procedure takes place resembles a woodworking shop. The orthopedist uses drills, screwdrivers, screws, staples, nails, chisels, and other tools to work the bone and connect pieces with each other.

At present, virtually any bone deformity can be corrected. Facial bones that are malformed can be reshaped or replaced. Bone transplants from one individual to another are commonplace. A patient who loses a limb from a disease such as cancer can have a normal-appearing prosthesis fitted and can be taught to use it to lead a near-normal lifestyle.

Orthopedists are also trained to treat several degenerative diseases such as arthritis, osteoporosis, carpal tunnel syndrome, and epicondylitis (tennis elbow). Treatment options may vary from diet changes to medications to steroid injections to exercise. Surgical procedures and hormone replacement therapy are additional options.

Recent technological advances such as joint replacement and the arthroscope (a specially designed illuminated surgical instrument) have benefitted orthopedic patients. Many orthopedic surgical procedures no longer require an open incision to expose the joint fully. Now, flexible arthroscopes can be inserted through a small incision in the skin and then into a joint, such as the knee, and then can be manipulated through the joint to locate and identify the nature of the injury. Arthroscopy can be used to look into many joints of the body. These include knees, shoulders, ankles, wrists, and elbows.

[See also **Skeletal system**]

Osmosis

Osmosis is the movement of a solvent, such as water, through a semipermeable membrane. (A solvent is the major component of a solution, the liquid in which something else is dissolved.) A semipermeable membrane is a material that allows some materials to flow through it but not others. The reason that semipermeable membranes have this property is that they contain very small holes. Small molecules, such as those of water, can flow easily through the holes. But large molecules, such as those of solutes (the component being dissolved, for instance sugar), cannot. Figure 1 illustrates this process. Notice that smaller molecules of water are able to pass through the openings in the membrane shown here but larger molecules of sugar are not.

Osmotic pressure

Osmosis always moves a solvent in one direction only, from a less concentrated solution to a more concentrated solution. As osmosis proceeds, pressure builds up on the side of the membrane where volume has increased. Ultimately, this pressure prevents more water from entering (for example, the bag in Figure 1), and osmosis stops. The osmotic pressure of a solution is the pressure needed to prevent osmosis from occurring.

Osmosis in living organisms

Living cells may be thought of as very small bags of solutions contained within semipermeable membranes. For example, Figure 1 might be thought of as a cell surrounded by a watery fluid. For the cell to survive, the concentration of substances within the cell must stay within a safe range.

A cell placed in a solution more concentrated than itself (a hypertonic solution) will shrink due to loss of water. It may eventually die of dehydration. You can observe this effect with a carrot placed in salty water. Within a few hours the carrot becomes limp and soft because its cells have shrivelled.

By contrast, a cell placed in a solution more dilute than itself (a hypotonic solution) will expand as water enters it. Under such conditions

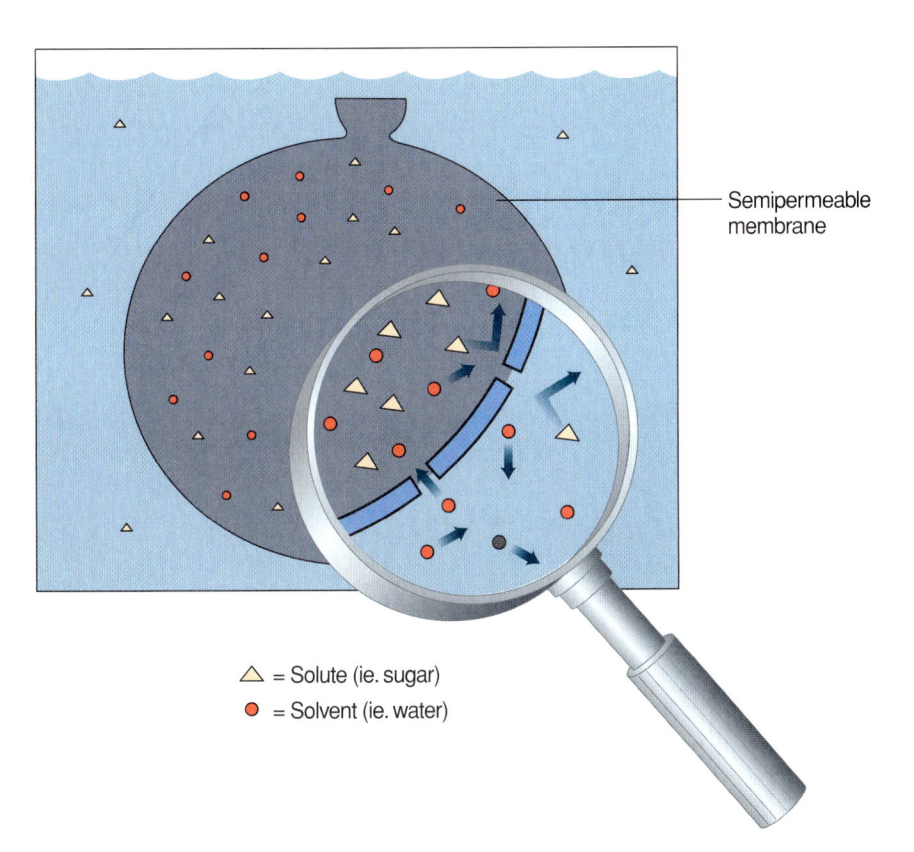

△ = Solute (ie. sugar)
● = Solvent (ie. water)

Figure 1. Osmosis is the movement of a solvent through a semipermeable membrane. *(Reproduced by permission of The Gale Group.)*

Osmosis

Words to Know

Concentration: The quantity of solute (for example sugar) dissolved in a given volume of solution (for example water).

Hypertonic solution: A solution with a higher osmotic pressure (solute concentration) than another solution.

Hypotonic solution: A solution with a lower osmotic pressure (solute concentration) than another solution.

Isotonic solutions: Two solutions that have the same concentration of solute particles and therefore the same osmotic pressure.

Osmotic pressure: The pressure which, applied to a solution in contact with pure solvent through a semipermeable membrane, will prevent osmosis from occurring.

Semipermeable membrane: A thin barrier between two solutions that permits only certain components of the solutions, usually the solvent, to pass through.

Solute: A substance dissolved to make a solution, for example sugar in sugar water.

Solution: A mixture of two or more substances that appears to be uniform throughout except on a molecular level.

Solvent: The major component of a solution or the liquid in which some other component is dissolved, for example water in sugar water.

---

the cell may burst. In general, plant cells are protected from bursting by the rigid cell wall that surrounds the cell membrane. As water enters the cell, it expands until it pushes up tight against the cell wall. The cell wall pushes back with an equal pressure, so no more water can enter.

Osmosis contributes to the movement of water through plants. Solute concentrations (the ratio of solutes to solvents in a solution) increase going from soil to root cells to leaf cells. The resulting differences of osmotic pressure help to push water upward. Osmosis also controls the evaporation of water from leaves by regulating the size of the openings (stomata) in the leaves' surfaces.

Organisms have various other methods for keeping their solute levels within safe range. Some cells live only in surroundings that are isotonic (have the same solute concentration as their own cells). For example, jellyfish that live in salt water have much higher salt-to-water solute concentrations than do freshwater creatures. Other animals continually replace lost water and solutes by drinking and eating. They remove excess water and solutes through excretion of urine.

Applications of osmosis

Preserving food. For thousands of years, perishable foods such as fish, olives, and vegetables have been preserved in salt or brine. The high salt concentration is hypertonic to bacteria cells, and kills them by dehydration before they can cause the food to spoil. Preserving fruit in sugar (as in jams or jellies) works on the same principle.

Artificial kidneys. People with kidney disease rely upon artificial kidney machines to remove waste products from their blood. Such machines use a process called dialysis, which is similar to osmosis. The difference between osmosis and dialysis is that a dialyzing membrane permits not just water, but also salts and other small molecules dissolved in the blood, to pass through. These materials move out of blood into a surrounding tank of distilled water. Red blood cells are too large to pass through the dialyzing membrane, so they return to the patient's body.

Desalination by reverse osmosis. Oceans hold about 97 percent of Earth's water supply, but their high salt content makes them unusable for drinking or agriculture. Salt can be removed by placing seawater in contact with a semipermeable membrane, then subjecting it to great pressure. Under these conditions, reverse osmosis occurs, by which pressure is used to push water from a more concentrated solution to a less concentrated solution. The process is just the reverse of the normal process of osmosis. In desalination, reverse osmosis is used to push water molecules out of seawater into a reservoir of pure water.

[*See also* **Diffusion; Solution**]

Oxidation-reduction reaction

The term oxidation-reduction reaction actually refers to two chemical reactions that always occur at the same time: oxidation and reduction. Oxidation-reduction reactions are also referred to more simply as redox

Oxidation-reduction reaction

> **Words to Know**
>
> **Combustion:** An oxidation-reduction reaction that occurs so rapidly that noticeable heat and light are produced.
>
> **Corrosion:** An oxidation-reduction reaction in which a metal is oxidized and oxygen is reduced, usually in the presence of moisture.
>
> **Oxidation:** A process in which a chemical substance takes on oxygen or loses electrons.
>
> **Oxidizing agent:** A chemical substance that gives up oxygen or takes on electrons from another substance.
>
> **Reducing agent:** A chemical substance that takes on oxygen or gives up electrons to another substance.
>
> **Reduction:** A process in which a chemical substance gives off oxygen or takes on electrons.

reactions. Oxidation, reduction, and redox reactions can all be defined in two ways.

The simpler definitions refer to reactions involving some form of oxygen. As an example, pure iron can be produced from iron oxide in a blast furnace by the following reaction:

$$3\ C + 2\ Fe_2O_3 \rightarrow 4\ Fe + 3\ CO_2$$

In this reaction, iron oxide (Fe_2O_3) gives away its oxygen to carbon (C). In chemical terms, the carbon is said to be oxidized because it has gained oxygen. At the same time, the iron oxide is said to be reduced because it has lost oxygen.

Because of its ability to give away oxygen, iron oxide is called an oxidizing agent. Similarly, because of its ability to take on oxygen, carbon is said to be a reducing agent. Oxidation and reduction always occur together. If one substance gives away oxygen (oxidation), a second substance must be present to take on that oxygen (reduction).

By looking at the above example, you can see that the following statements must always be true:

An oxidizing agent (in this case, iron oxide) is always reduced.

A reducing agent (in this case, carbon) is always oxidized.

Redox and electron exchanges

For many years, chemists thought of oxidation and reduction as involving the element oxygen in some way or another. That's where the name oxidation came from. But they eventually learned that other elements behave chemically in much the same way as oxygen. They decided to revise their definition of oxidation and reduction to make it more general—to apply to elements other than oxygen.

The second definition for oxidation and reduction is not as easy to see. It is based on the fact that when two elements react with each other, they do so by exchanging electrons. In an oxidation-reduction reaction like the one above, the element that is oxidized always loses electrons. The element that is reduced always gains electrons. The more general definition of redox reactions, then, involves the gain and loss of electrons rather than the gain and loss of oxygen.

In the reaction below, for example, sodium metal (Na) reacts with chlorine gas (Cl_2) in such a way that sodium atoms lose one electron each to chlorine atoms:

$$2\, Na + Cl_2 \rightarrow 2\, NaCl$$

Because sodium loses electrons in this reaction, it is said to be oxidized. Because chlorine gains electrons in the reaction, it is said to be reduced.

Types of redox reactions. Redox reactions are among the most common and most important chemical reactions in everyday life. The great majority of those reactions can be classified on the basis of how rapidly they occur. Combustion is an example of a redox reaction that occurs so rapidly that noticeable heat and light are produced. Corrosion, decay, and various biological processes are examples of oxidation that occurs so slowly that noticeable heat and light are *not* produced.

Combustion. Combustion means burning. Any time a material burns, an oxidation-reduction reaction occurs. The two equations below show what happens when coal (which is nearly pure carbon) and gasoline (C_8H_{18}) burn. You can see that the fuel is oxidized in each case:

$$C + O_2 \rightarrow CO_2$$
$$2\, C_8H_{18} + 25\, O_2 \rightarrow 16\, CO_2 + 18\, H_2O$$

In reactions such as these, oxidation occurs very rapidly and energy is released. That energy is put to use to heat homes and buildings; to drive automobiles, trucks, ships, airplanes, and trains; to operate industrial processes; and for numerous other purposes.

Rust. Most metals react with oxygen to form compounds known as oxides. Rust is the name given to the oxide of iron and, sometimes, the oxides of other metals. The process by which rusting occurs is also known as corrosion. Corrosion is very much like combustion, except that it occurs much more slowly. The equation below shows perhaps the most common form of corrosion, the rusting of iron.

$$4\,Fe + 3\,O_2 \rightarrow 2\,Fe_2O_3$$

Decay. The compounds that make up living organisms, such as plants and animals, are very complex. They consist primarily of carbon, oxygen, and hydrogen. A simple way to represent such compounds is to use the letters x, y, and z to show that many atoms of carbon, hydrogen, and oxygen are present in the compounds.

When a plant or animal dies, the organic compounds of which it is composed begin to react with oxygen. The reaction is similar to the combustion of gasoline shown above, but it occurs much more slowly. The process is known as decay, and it is another example of a common oxidation-reduction reaction. The equation below represents the decay (oxidation) of a compound that might be found in a dead plant:

$$C_xH_yO_z + O_2 \rightarrow CO_2 + H_2O$$

Biological processes. Many of the changes that take place within living organisms are also redox reactions. For example, the digestion of food is an oxidation process. Food molecules react with oxygen in the body to form carbon dioxide and water. Energy is also released in the process. The carbon dioxide and water are eliminated from the body as waste products, but the energy is used to make possible all the chemical reactions that keep an organism alive and help it to grow.

Oxygen family

The oxygen family consists of the elements that make up group 16 on the periodic table: oxygen, sulfur, selenium, tellurium, and polonium. These elements all have six electrons in their outermost energy level, accounting for some common chemical properties among them. In another respect, the elements are quite different from each other. Oxygen is a gaseous nonmetal; sulfur and selenium are solid nonmetals; tellurium is a solid metalloid; and polonium is a solid metal.

Words to Know

Acid: Substances that, when dissolved in water, are capable of reacting with a base to form salts and release hydrogen ions.

Allotrope: One of two or more forms of an element.

Combustion: A form of oxidation that occurs so rapidly that noticeable heat and light are produced.

Cracking: The process by which large hydrocarbon molecules are broken down into smaller components.

Electrolysis: The process by which an electrical current causes a chemical change, usually the breakdown of some substance.

Isotopes: Two or more forms of the same element with the same number of protons but different numbers of neutrons in the atomic nucleus.

Lithosphere: The solid portion of Earth, especially the outer crustal region.

LOX: An abbreviation for liquid oxygen.

Metallurgy: The science and technology that deals with gaining metals from their ores and converting them into forms that have practical value.

Nascent oxygen: An allotrope of oxygen whose molecules each contain a single oxygen atom.

Ozone: An allotrope of oxygen that consists of three atoms per molecule.

Producer gas: A synthetic fuel that consists primarily of carbon monoxide and hydrogen gases.

Proteins: Large molecules that are essential to the structure and functioning of all living cells.

Radioactive decay: The predictable manner in which a population of atoms of a radioactive element spontaneously disintegrate over time.

Oxygen

Oxygen is a colorless, odorless, tasteless gas with a melting point of −218°C (−360°F) and a boiling point of −183°C (−297°F). It is the most abundant element in Earth's crust, making up about one-quarter of the atmosphere by weight, about one-half of the lithosphere (Earth's

Oxygen family

crust), and about 85 percent of the hydrosphere (the oceans, lakes, and other forms of water). It occurs both as a free element and in a large variety of compounds. In the atmosphere, it exists as elemental oxygen, sometimes known as dioxygen because it consists of diatomic molecules, O_2. In water it occurs as hydrogen oxide, H_2O, and in the lithosphere it occurs in compounds such as oxides, carbonates, sulfates, silicates, phosphates, and nitrates.

Oxygen also exists in two allotropic forms (physically or chemically different forms of the same substance): one atom per molecule (O) and three atoms per molecule (O_3). The former allotrope is known as monatomic, or nascent, oxygen and the latter as triatomic oxygen, or ozone. Under most circumstances in nature, the diatomic form of oxygen predominates. In the upper part of the stratosphere, however, solar energy causes the breakdown of the diatomic form into the monatomic form, which may then recombine with diatomic molecules to form ozone. The presence of ozone in Earth's atmosphere is critical for the survival of life on Earth since that allotrope has a tendency to absorb ultraviolet radiation that would otherwise be harmful or even fatal to both plant and animal life on the planet's surface.

Oxygen was discovered independently by Swedish chemist Carl Scheele (1742–1786) and English chemist Joseph Priestley (1733–1804) in the period between 1773 and 1774. The element was given its name in the late 1770s by French chemist Antoine Laurent Lavoisier (1743–1794). Its name comes from the French word for "acid-former," reflecting Lavoisier's incorrect belief that all acids contain oxygen.

Production. By far the most common method for producing oxygen commercially is by the fractional distillation of liquid air. A sample of air is first cooled to a very low temperature in the range of $-200°C$ ($-330°F$). At this temperature, most gases that make up air become liquid. The liquid air is then allowed to evaporate. At a temperature of about $-196°C$ ($-320°F$), nitrogen begins to boil off. When most of the nitrogen is gone, argon and neon also boil off, leaving an impure form of oxygen behind. The oxygen is impure because small amounts of krypton, xenon, and other gases may remain in the liquid form. In order to further purify the oxygen, the process of cooling, liquefying, and evaporation may be repeated.

Oxygen is commonly stored and transported in its liquid form, a form also known as LOX (for *l*iquid *ox*ygen). LOX containers look like very large vacuum bottles consisting of a double-walled container with a vacuum between the walls. The element can also be stored and transported less easily in gaseous form in steel-walled containers about 1.2 meters (4 feet) high and 23 centimeters (9 inches) in diameter. In many instances, oxygen is manufactured at the location where it will be used. The process of fractional distillation described earlier is sufficiently simple and inexpensive so that many industries can provide their own oxygen-production facilities.

Uses. Oxygen has so many commercial, industrial, and other uses that it consistently ranks among the top five chemicals in volume of production in the United States. In 1990, for example, about 18 billion kilograms (39 billion pounds) of the element were manufactured in the United States.

The uses to which oxygen is put can be classified into four major categories: metallurgy, rocketry, chemical synthesis, and medicine. In the processing of iron ore in a blast furnace, for example, oxygen is used to convert coke (carbon) to carbon monoxide. The carbon monoxide, in turn, reduces iron oxides to pure iron metal. Oxygen is then used in a second step of iron processing in the Bessemer converter, open hearth, or basic oxygen process method of converting "pig iron" to steel. In this step, the oxygen is used to react with the excess carbon, silicon, and metals remaining in the pig iron that must be removed in order to produce steel.

Another metallurgical application of oxygen is in torches used for welding and cutting. The two most common torches make use of the reaction between oxygen and hydrogen (the oxyhydrogen torch) or between oxygen and acetylene (the oxyacetylene torch). Both kinds of torch produce temperatures in the range of 3,000°C (5,400°F) or more and can, therefore, be used to cut through or weld the great majority of metallic materials.

In the form of LOX, oxygen is used widely as the oxidizing agent in many kinds of rockets and missiles. For example, the huge external fuel tank required to lift the space shuttle into space holds 550,000 liters (145,000 gallons) of liquid oxygen and 1,500,000 liters (390,000 gallons) of liquid hydrogen. When these two elements react in the shuttle's main engines, they provide a maximum thrust of 512,000 pounds.

The chemical industry uses vast amounts of oxygen every year in a variety of chemical synthesis (formation) reactions. One of the most important of these is the cracking of hydrocarbons by oxygen. Under most circumstances, heating a hydrocarbon with oxygen results in combustion, with carbon dioxide and water as the main products. However, if the rate at which oxygen is fed into a hydrocarbon mixture is carefully controlled, the hydrocarbon is "cracked," or broken apart to produce other products, such as acetylene, ethylene, and propylene.

Various types of synthetic fuels can also be manufactured with oxygen as one of the main reactants. Producer gas, as an example, is manufactured by passing oxygen at a controlled rate through a bed of hot coal or coke. The majority of carbon dioxide produced in this reaction is reduced to carbon monoxide so that the final product (the producer gas) consists primarily of carbon monoxide and hydrogen.

Perhaps the best-known medical application of oxygen is in oxygen therapy, where patients who are having trouble breathing are given doses of pure or nearly pure oxygen. Oxygen therapy is often used during surgical procedures, during childbirth, during recovery from heart attacks, and during treatment for infectious diseases. In each case, providing a person with pure oxygen reduces the stress on his or her heart and lungs, speeding the rate of recovery.

Pure oxygen or air enriched with oxygen may also be provided in environments where breathing may be difficult. Aircraft that fly at high altitudes, of course, are always provided with supplies of oxygen in case of any problems with the ship's normal air supply. Deep-sea divers also carry with them or have pumped to them supplies of air that are enriched with oxygen.

Some water purification and sewage treatment plants use oxygen. The gas is pumped through water to increase the rate at which naturally occurring bacteria break down organic waste materials. A similar process has been found to reduce the rate at which eutrophication takes place in lakes and ponds and, in some cases, to actually reverse that process. (Eutrophication is the dissolving of nutrients in a body of water. Growth in aquatic plant life and a decrease in dissolved oxygen are the two main results of the process.)

Finally, oxygen is essential to all animal life on Earth. A person can survive a few days or weeks without water or food but no more than a few minutes without oxygen. In the absence of oxygen, energy-generating chemical reactions taking place within cells would come to an end, and a person would die.

Sulfur

Sulfur is a nonmetallic element that can exist in many allotropic forms (physically or chemically different forms of the same substance). The most familiar are called rhombic and monoclinic sulfur. Both are bright yellow solids with melting points of about 115°C (239°F). A third form is called plastic or amorphous sulfur. It is a brownish liquid produced when rhombic or monoclinic sulfur is melted.

Sulfur itself has no odor at all. It has a bad reputation in this regard, however, because some of its most common compounds have strong smells. Sulfur dioxide, one of these compounds, has a sharp, choking, suffocating effect on anyone who breathes it. The "fire and brimstone" of the Bible was one of the worst punishments that its authors could imagine. The brimstone in this expression referred to burning sulfur, or sulfur dioxide. The fact that sulfur comes from deep under the ground and that sulfur dioxide can be smelled in the fumes of volcanoes further fueled people's imaginations of what Hell must be like.

A second sulfur compound with a bad odor is hydrogen sulfide. The strong smell of rotten eggs is due to the presence of this compound.

Occurrence and preparation. Sulfur is the sixteenth most abundant element in Earth's crust. It occurs both as an element and in a variety of compounds. As an element it can be found in very large, underground mines, most commonly along the Gulf Coast of the United States and in Poland and Sicily. The sulfur is extracted from these mines by means of the Frasch process. In this process, superheated steam is pumped through the outermost of a set of three pipes. Compressed air is forced down the innermost pipe. The superheated steam causes the underground sulfur to melt, and the compressed air forces it upward, through the middle of the three pipes, to Earth's surface.

Sulfur is also widely distributed in the form of minerals and ores. Many of these are in the form of sulfates, including gypsum (calcium sulfate, $CaSO_4$), barite (barium sulfate, $BaSO_4$), and Epsom salts (magnesium sulfate, $MgSO_4$). Others are metal sulfides, including iron pyrites (iron sulfide, FeS_2), galena (lead sulfide, PbS), cinnabar (mercuric sulfide, HgS), stibnite (antimony sulfide, Sb_2S_3), and zinc blende (zinc sulfide, ZnS). The sulfur is recovered from these metal ores by heating them strongly in air, which converts the sulfur to sulfur dioxide and releases the pure metal. Then the sulfur dioxide can go directly into the manufacture of sulfuric acid, which is where more than 90 percent of the world's mined sulfur winds up.

Uses of sulfur and its compounds. Some sulfur is used directly as a fungicide and insecticide, in matches, fireworks, and gunpowder, and in the vulcanization of natural rubber (a treatment that gives rubber elasticity and strength). Most, however, is converted into a multitude of useful compounds.

Sulfuric acid is by far the most important of all sulfur compounds. Nearly 90 percent of all sulfur produced is converted first into sulfur dioxide and then into sulfuric acid. The acid consistently ranks number one among the chemicals produced in the United States. In 1990, more than 40 billion kilograms (89 billion pounds) of sulfuric acid were

Oxygen family

manufactured, more than 50 percent as much as the second most popular chemical (nitrogen gas). Sulfuric acid is used in the production of fertilizers, automobile batteries, petroleum products, pigments, iron and steel, and many other products.

The sulfur cycle. Like nitrogen, carbon, and phosphorus, sulfur passes through the gaseous, liquid, and solid parts of our planet in a series of continuous reactions known as the sulfur cycle. The main steps in the sulfur cycle are illustrated in the accompanying figure.

Sulfur is produced naturally as a result of volcanic eruptions and through emissions from hot springs. It enters the atmosphere primarily in the form of sulfur dioxide, then remains in the atmosphere in that form or, after reacting with water, in the form of sulfuric acid. Sulfur is carried back to Earth's surface as acid deposition when it rains or snows.

The sulfur cycle. *(Reproduced by permission of The Gale Group.)*

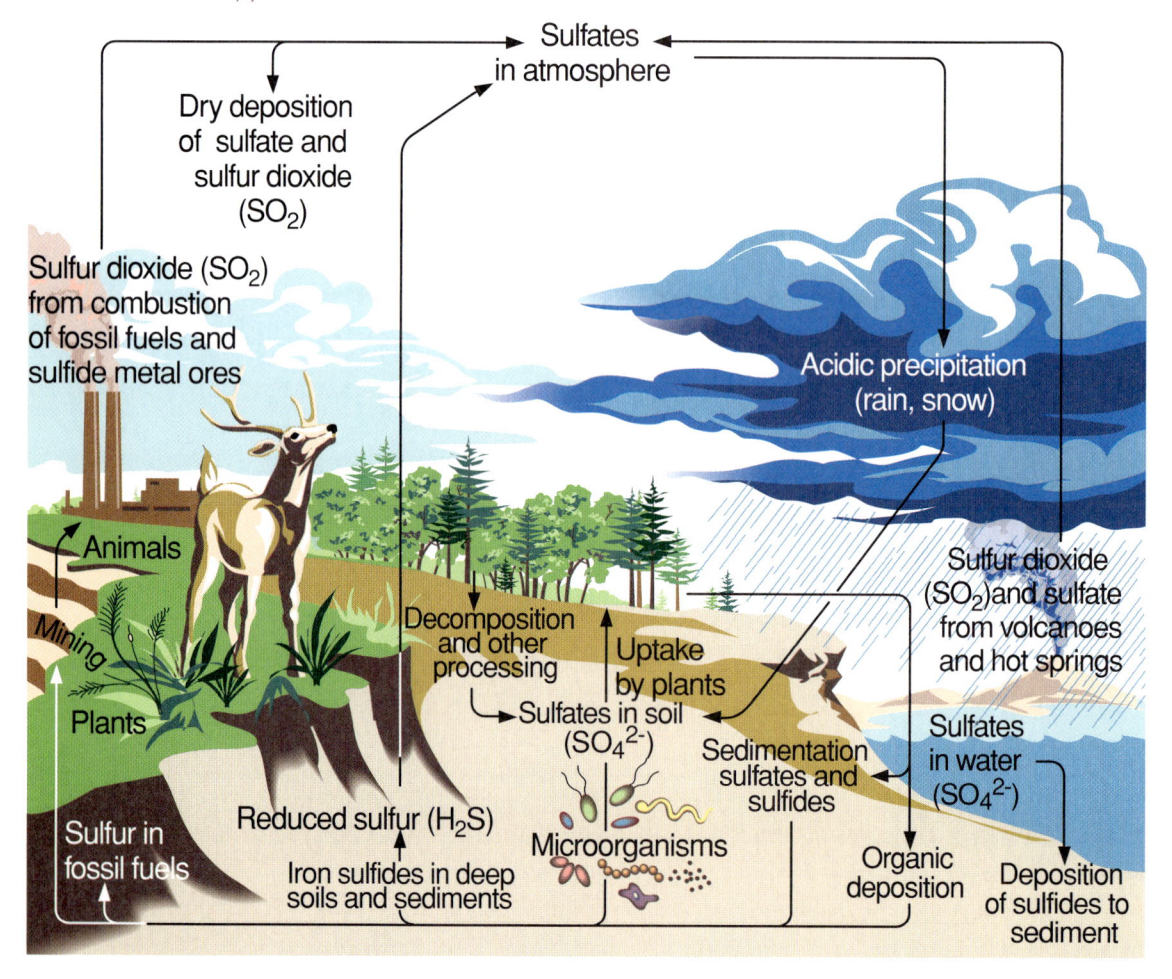

On Earth's surface, sulfur dioxide and sulfuric acid react with metals to form sulfates and sulfides. The element is also incorporated by plants in a form known as organic sulfur. Certain amino acids, the compounds from which proteins are made, contain sulfur. Organic sulfur from plants is eventually passed on to animals that eat those plants. It is, in turn, converted from plant proteins to animal proteins.

When plants and animals die, sulfur is returned to the soil where it is converted by microorganisms into hydrogen sulfide. Hydrogen sulfide gas is then returned to the atmosphere, where it is oxidized to sulfuric acid.

Human activities influence the sulfur cycle in a number of ways. For example, when coal and metallic ores are mined, sulfur and sulfides may be released and returned to the soil. Also, the combustion of coal, oil, and natural gas often releases sulfur dioxide to the atmosphere. This sulfur dioxide is added to the amount already present from natural sources, greatly increasing the amount of acid precipitation that falls to Earth's surface. Some people believe that acid precipitation (or acid rain) is responsible for the death of trees and other plants, the acidification of lakes that has hurt marine animals, damage to metal and stone structures, and other environmental harm.

Selenium, tellurium, and polonium

Selenium and tellurium are both relatively rare elements. They rank in the bottom ten percent of all elements in terms of abundance. They tend to occur in Earth's crust in association with ores of copper and other metals. Both are obtained as a by-product of the electrolytic refining of copper. During that process, they sink to the bottom of the electrolysis tank, where they can be removed from the sludge that develops.

Selenium occurs in a variety of allotropic forms (physically or chemically different forms of the same substance), the most common of which is a red powder that becomes black when exposed to air. The element's melting point is 217°C (423°F), and its boiling point is 685°C (1,265°F). Tellurium is a silvery-white solid that looks like a metal (although it is actually a metalloid). Its melting point is 450°C (842°F), and its boiling point is 990°C (1,814°F).

Selenium has an interesting role in living organisms. It is essential in very low concentrations for maintaining health in most animals. In fact, it is often added to animal feeds. In higher concentrations, however, the element has been found to have harmful effects on animals, causing deformed young and diseased adults.

The primary uses of selenium are in electronics and in the manufacture of colored glass. Photocopying machinery, solar cells, photocells, television picture tubes, and electronic rectifiers and relays (used to control the flow of electric current) all use selenium. Some of the most beautiful colored glasses, ranging from pale pink to brilliant reds, are made with compounds of selenium.

Small amounts of tellurium are also used in the production of colored glass. More than 90 percent of the element, however, goes to the production of alloys of iron and other metals.

Polonium has 27 isotopes, all of which are radioactive. It occurs naturally in uranium ores, where it is the final product in the long series of reactions by which uranium undergoes radioactive decay. It is one of the rarest elements on Earth, with an abundance of no more than about 3×10^{-10} parts per million. The discovery of polonium in 1898 by Polish-French chemist Marie Curie (1867–1934) is one of the most dramatic stories in the history of science. She processed tons of uranium ore in order to obtain a few milligrams of the new element, which she then named after her homeland of Poland. Polonium finds limited use in highly specialized power-generating devices, such as those used for space satellites and space probes.

Ozone

Ozone is an allotrope (a physically or chemically different form of the same substance) of oxygen with the chemical formula O_3. This formula shows that each molecule of ozone consists of three atoms. By comparison, normal atmospheric oxygen—also known as dioxygen—consists of two atoms per molecule and has the chemical formula O_2.

Ozone is a bluish gas with a sharp odor that decomposes readily to produce dioxygen. Its normal boiling point is $-112°C$ ($-170°F$), and its freezing point is $-192°C$ ($-314°F$). It is more soluble in water than dioxygen and also much more reactive. Ozone occurs in the lower atmosphere in very low concentrations, but it is present in significantly higher concentrations in the upper atmosphere. The reason for this difference is that energy from the Sun causes the decomposition of oxygen molecules in the upper atmosphere:

$$O_2 \rightarrow (\text{solar energy}) \rightarrow 2O$$

The nascent (single-atom) oxygen formed is very reactive. It may combine with other molecules of dioxygen to form ozone:

$$O + O_2 \rightarrow O_3$$

Ozone

Words to Know

Allotropes: Forms of a chemical element with different physical and chemical properties.

Chlorofluorocarbons (CFCs): A family of chemical compounds consisting of carbon, fluorine, and chlorine.

Dioxygen: The name sometimes used for ordinary atmospheric oxygen with the chemical formula O_2.

Electromagnetic radiation: A form of energy carried by waves.

Nascent oxygen: Oxygen that consists of molecules made of a single oxygen atom, O.

Ozone hole: A term invented to describe a region of very low ozone concentration above the Antarctic that appears and disappears with each austral (Southern Hemisphere) summer.

Ozone layer: A region of the stratosphere in which the concentration of ozone is relatively high.

Radiation: Energy transmitted in the form of electromagnetic waves or subatomic particles.

Standard (pollution): The highest level of a harmful substance that can be present without a serious possibility of damaging plant or animal life.

Stratosphere: The region of Earth's atmosphere ranging between about 15 and 50 kilometers (9 and 30 miles) above Earth's surface.

Troposphere: The lowest layer of Earth's atmosphere, ranging to an altitude of about 15 kilometers (9 miles) above Earth's surface.

Ultraviolet radiation: A form of electromagnetic radiation with wavelengths just less than those of visible light (4 to 400 nanometers, or billionths of a meter).

Ozone layer depletion

Most of the ozone in our atmosphere is concentrated in a region of the stratosphere between 15 and 30 kilometers (9 and 18 miles) above Earth's surface. The total amount of ozone in this band is actually relatively small. If it were all transported to Earth's surface, it would form a

Ozone

layer no more than 3 millimeters (about 0.1 inch) thick. Yet stratospheric ozone serves an invaluable function to life on Earth.

Radiation from the Sun that reaches Earth's outer atmosphere consists of a whole range of electromagnetic radiation: cosmic rays, gamma rays, ultraviolet radiation, infrared radiation, and visible light. Various forms of radiation can have both beneficial and harmful effects. Ultraviolet radiation, for example, is known to affect the growth of certain kinds of plants, to cause eye damage in animals, to disrupt the function of DNA (the genetic material in an organism), and to cause skin cancer in humans.

Fortunately for living things on Earth, ozone molecules absorb radiation in the ultraviolet region. Thus, the ozone layer in the stratosphere protects plants and animals on Earth's surface from most of these dangerous effects.

Human effects on the ozone layer. In 1984, scientists reported that the ozone layer above the Antarctic appeared to be thinning. In fact, the amount of ozone dropped to such a low level that the term "hole" was used to describe the condition. The hole was a circular area above the Antarctic in which ozone had virtually disappeared. In succeeding years, that hole reappeared with the onset of each summer season in the Antarctic (September through December).

The potential threat to humans (and other organisms) was obvious. Increased exposure to ultraviolet radiation because of a thinner ozone layer would almost certainly mean higher rates of skin cancer. Other medical problems were also possible.

An atmosphere with a relatively intact ozone layer (left) compared to one with an ozone layer damaged by chlorofluorocarbon (CFC) emissions. *(Reproduced by permission of The Gale Group.)*

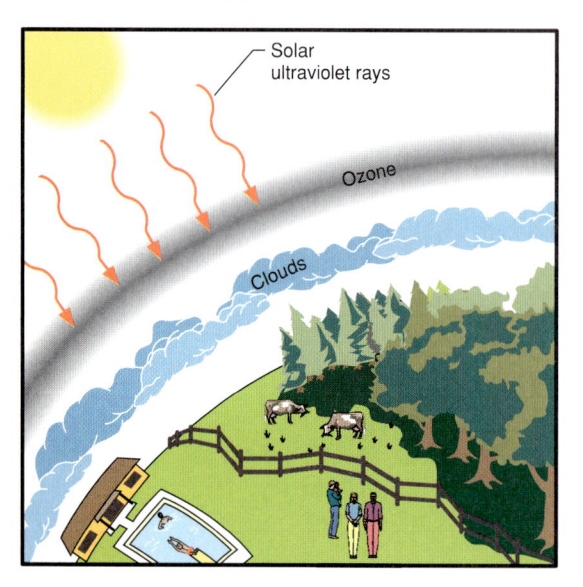

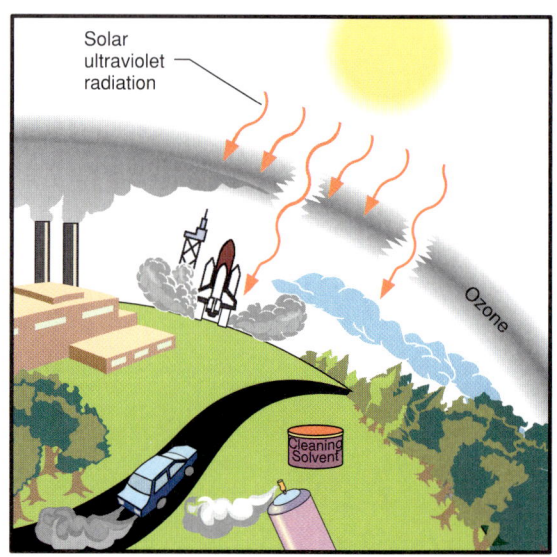

At first, scientists disagreed as to the cause of the thinning ozone layer. Eventually, however, the evidence seemed to suggest that chemicals produced and made by humans might be causing the destruction of the ozone. In particular, a group of compounds known as the chlorofluorocarbons (CFCs) were suspected. These compounds had become widely popular in the 1970s and 1980s for a number of applications, including as chemicals used in refrigeration, as propellants in aerosol sprays, as blowing agents in the manufacture of plastic foams and insulation, as dry-cleaning fluids, and as cleaning agents for electronic components.

One reason for the popularity of the CFCs was their stability. They normally do not break down when used on Earth's surface. In the upper atmosphere, however, the situation changes. Evidence suggests that CFCs break down to release chlorine atoms which, in turn, attack and destroy ozone molecules:

$$CFC \rightarrow \text{solar energy} \rightarrow Cl \text{ atoms}$$

$$Cl + O_3 \rightarrow ClO + O_2$$

This process is especially troublesome because one of the products of the reaction, chlorine monoxide (ClO) reacts with other molecules of the same kind to generate more chlorine atoms:

$$ClO + ClO \rightarrow Cl + Cl + O_2$$

Once CFCs get into the stratosphere and break down, therefore, a continuous supply of chlorine atoms is assured. And those chlorine atoms destroy ozone molecules.

Scientists and nonscientists alike soon became concerned about the role of CFCs in the depletion of stratospheric ozone. A movement then developed to reduce and/or ban the use of these chemicals. In 1987, a conference sponsored by the United Nations Environment Programme resulted in the so-called Montreal Protocol. The Protocol set specific time limits for the phasing out of both the production and use of CFCs. Only three years later, concern had become so great that the Protocol deadlines were actually moved up. One hundred and sixty-five nations signed this agreement. Because of the Protocol, the United States, Australia, and other developed countries have completely phased out the production of CFCs. According to the Protocol, developing nations have until the year 2010 to complete their phase out.

Ozone in the troposphere

Ozone is a classic example of a chemical that is both helpful and harmful. In the stratosphere, of course, it is essential in protecting plants

Ozone

and animals on Earth's surface from damage by ultraviolet radiation. But in the lower regions of the atmosphere, near Earth's surface, the story is very different.

The primary source of ozone on Earth is the internal-combustion engine. Gases released from the tailpipe of a car or truck can be oxidized in the presence of sunlight to produce ozone. Ozone itself has harmful effects on both plants and animals. In humans and other animals, the gas irritates and damages membranes of the respiratory system and eyes. It can also induce asthma. Sensitive people are affected at concentrations that commonly occur on an average city street during rush-hour traffic.

Ozone exposure also brings on substantial damage to both agricultural and wild plants. Its primary effect is to produce a distinctive injury that reduces the area of foliage on which photosynthesis can occur. (Photosynthesis is a complicated process in which plants utilize light energy to form carbohydrates and release oxygen as a by-product.) Most plants are seriously injured by a two- to four-hour exposure to high levels of ozone. But long-term exposures to even low levels of the gas can cause decreases in growth. Large differences among plants exist, with tobacco, spinach, and conifer trees being especially sensitive.

Many nations, states, and cities have now set standards for maximum permissible concentrations of ozone in their air. At the present time in the United States, the standard is 120 ppb (parts per billion). That number had been raised from 80 ppb in 1979 because many urban areas could not meet the lower standard. Areas in which ozone pollution is most severe—such as Los Angeles, California—cannot meet even the higher standard. Measurements of 500 ppb for periods of one hour in Los Angeles are not uncommon.

Long-term problem

Despite a relatively rapid and effective international response to CFC emissions, the recovery of the ozone layer may take up to 50 years or more. This is because these chemicals are very persistent in the environment: CFCs already present will also be around for many decades. Moreover, there will continue to be substantial emissions of CFCs for years after their manufacture, and uses are banned because older CFC-containing equipment and products already in use continue to release these chemicals.

In studies released in late 2000, scientists said they were stunned by findings that up to 70 percent of the ozone layer over the North Pole has been lost and that the ozone hole over the South Pole grew to an expanse larger than North America. According to the National Oceanic and Atmospheric Administration (NOAA), the hole in the ozone layer over the South Pole expanded to a record 17.1 million square miles (44.3 million square kilometers).

Scientists blamed the record ozone holes on two main reasons: extreme cold and the continued use of bromine. Recent very cold winters in the two poles have slowed the recovery of the ozone layer. Cold air slows the dissipation and decay of CFCs, which allows them to destroy ozone faster. Bromine is a chemical cousin to chlorine and is used for some of the same purposes—fire fighting, infection control, and sanitation. NOAA believes bromine is 45 times more damaging to ozone in the atmosphere than chlorine. But bromine has not been regulated as strictly as chlorine because countries could not stand the loss of income if it were regulated more. Some scientists, however, believe governments will be under growing pressure in the coming years to limit the chemical.

[*See also* **Greenhouse effect**]

Where to Learn More

Books

Earth Sciences

Cox, Reg, and Neil Morris. *The Natural World.* Philadelphia, PA: Chelsea House, 2000.

Dasch, E. Julius, editor. *Earth Sciences for Students.* Four volumes. New York: Macmillan Reference, 1999.

Denecke, Edward J., Jr. *Let's Review: Earth Science.* Second edition. Hauppauge, NY: Barron's, 2001.

Engelbert, Phillis. *Dangerous Planet: The Science of Natural Disasters.* Three volumes. Farmington Hills, MI: UXL, 2001.

Gardner, Robert. *Human Evolution.* New York: Franklin Watts, 1999.

Hall, Stephen. *Exploring the Oceans.* Milwaukee, WI: Gareth Stevens, 2000.

Knapp, Brian. *Earth Science: Discovering the Secrets of the Earth.* Eight volumes. Danbury, CT: Grolier Educational, 2000.

Llewellyn, Claire. *Our Planet Earth.* New York: Scholastic Reference, 1997.

Moloney, Norah. *The Young Oxford Book of Archaeology.* New York: Oxford University Press, 1997.

Nardo, Don. *Origin of Species: Darwin's Theory of Evolution.* San Diego, CA: Lucent Books, 2001.

Silverstein, Alvin, Virginia Silverstein, and Laura Silverstein Nunn.*Weather and Climate.* Brookfield, CN: Twenty-First Century Books, 1998.

Williams, Bob, Bob Ashley, Larry Underwood, and Jack Herschbach. *Geography.* Parsippany, NJ: Dale Seymour Publications, 1997.

Life Sciences

Barrett, Paul M. *National Geographic Dinosaurs.* Washington, D.C.: National Geographic Society, 2001.

Fullick, Ann. *The Living World.* Des Plaines, IL: Heinemann Library, 1999.

Gamlin, Linda. *Eyewitness: Evolution.* New York: Dorling Kindersley, 2000.

Greenaway, Theresa. *The Plant Kingdom: A Guide to Plant Classification and Biodiversity.* Austin, TX: Raintree Steck-Vaughn, 2000.

Kidd, J. S., and Renee A Kidd. *Life Lines: The Story of the New Genetics.* New York: Facts on File, 1999.

Kinney, Karin, editor. *Our Environment.* Alexandria, VA: Time-Life Books, 2000.

Where to Learn More

Nagel, Rob. *Body by Design: From the Digestive System to the Skeleton.* Two volumes. Farmington Hills, MI: UXL., 2000.

Parker, Steve. *The Beginner's Guide to Animal Autopsy: A "Hands-in" Approach to Zoology, the World of Creatures and What's Inside Them.* Brookfield, CN: Copper Beech Books, 1997.

Pringle, Laurence. *Global Warming: The Threat of Earth's Changing Climate.* New York: SeaStar Books, 2001.

Riley, Peter. *Plant Life.* New York: Franklin Watts, 1999.

Stanley, Debbie. *Genetic Engineering: The Cloning Debate.* New York: Rosen Publishing Group, 2000.

Whyman, Kate. *The Animal Kingdom: A Guide to Vertebrate Classification and Biodiversity.* Austin, TX: Raintree Steck-Vaughn, 1999.

Physical Sciences

Allen, Jerry, and Georgiana Allen. *The Horse and the Iron Ball: A Journey Through Time, Space, and Technology.* Minneapolis, MN: Lerner Publications, 2000.

Berger, Samantha, *Light.* New York: Scholastic, 1999.

Bonnet, Bob L., and Dan Keen. *Physics.* New York: Sterling Publishing, 1999.

Clark, Stuart. *Discovering the Universe.* Milwaukee, WI: Gareth Stevens, 2000.

Fleisher, Paul, and Tim Seeley. *Matter and Energy: Basic Principles of Matter and Thermodynamics.* Minneapolis, MN: Lerner Publishing, 2001.

Gribbin, John. *Eyewitness: Time and Space.* New York: Dorling Kindersley, 2000.

Holland, Simon. *Space.* New York: Dorling Kindersley, 2001.

Kidd, J. S., and Renee A. Kidd. *Quarks and Sparks: The Story of Nuclear Power.* New York: Facts on File, 1999.

Levine, Shar, and Leslie Johnstone. *The Science of Sound and Music.* New York: Sterling Publishing, 2000

Naeye, Robert. *Signals from Space: The Chandra X-ray Observatory.* Austin, TX: Raintree Steck-Vaughn, 2001.

Newmark, Ann. *Chemistry.* New York: Dorling Kindersley, 1999.

Oxlade, Chris. *Acids and Bases.* Chicago, IL: Heinemann Library, 2001.

Vogt, Gregory L. *Deep Space Astronomy.* Brookfield, CT: Twenty-First Century Books, 1999.

Technology and Engineering Sciences

Baker, Christopher W. *Scientific Visualization: The New Eyes of Science.* Brookfield, CT: Millbrook Press, 2000.

Cobb, Allan B. *Scientifically Engineered Foods: The Debate over What's on Your Plate.* New York: Rosen Publishing Group, 2000.

Cole, Michael D. *Space Launch Disaster: When Liftoff Goes Wrong.* Springfield, NJ: Enslow, 2000.

Deedrick, Tami. *The Internet.* Austin, TX: Raintree Steck-Vaughn, 2001.

DuTemple, Leslie A. *Oil Spills.* San Diego, CA: Lucent Books, 1999.

Gaines, Ann Graham. *Satellite Communication.* Mankata, MN: Smart Apple Media, 2000.

Gardner, Robert, and Dennis Shortelle. *From Talking Drums to the Internet: An Encyclopedia of Communications Technology.* Santa Barbara, CA: ABC-Clio, 1997.

Graham, Ian S. *Radio and Television.* Austin, TX: Raintree Steck-Vaughn, 2000.

Parker, Steve. *Lasers: Now and into the Future.* Englewood Cliffs, NJ: Silver Burdett Press, 1998.

Sachs, Jessica Snyder. *The Encyclopedia of Inventions.* New York: Franklin Watts, 2001.

Wilkinson, Philip. *Building.* New York: Dorling Kindersley, 2000.

Wilson, Anthony. *Communications: How the Future Began.* New York: Larousse Kingfisher Chambers, 1999.

Periodicals

Archaeology. Published by Archaeological Institute of America, 656 Beacon Street, 4th Floor, Boston, Massachusetts 02215. Also online at www.archaeology.org.

Astronomy. Published by Kalmbach Publishing Company, 21027 Crossroads Circle, Brookfield, WI 53186. Also online at www.astronomy.com.

Discover. Published by Walt Disney Magazine, Publishing Group, 500 S. Buena Vista, Burbank, CA 91521. Also online at www.discover.com.

National Geographic. Published by National Geographic Society, 17th & M Streets, NW, Washington, DC 20036. Also online at www.nationalgeographic.com.

New Scientist. Published by New Scientist, 151 Wardour St., London, England W1F 8WE. Also online at www.newscientist.com (includes links to more than 1,600 science sites).

Popular Science. Published by Times Mirror Magazines, Inc., 2 Park Ave., New York, NY 10024. Also online at www.popsci.com.

Science. Published by American Association for the Advancement of Science, 1333 H Street, NW, Washington, DC 20005. Also online at www.sciencemag.org.

Science News. Published by Science Service, Inc., 1719 N Street, NW, Washington, DC 20036. Also online at www.sciencenews.org.

Scientific American. Published by Scientific American, Inc., 415 Madison Ave, New York, NY 10017. Also online at www.sciam.com.

Smithsonian. Published by Smithsonian Institution, Arts & Industries Bldg., 900 Jefferson Dr., Washington, DC 20560. Also online at www.smithsonianmag.com.

Weatherwise. Published by Heldref Publications, 1319 Eighteenth St., NW, Washington, DC 20036. Also online at www.weatherwise.org.

Web Sites

Cyber Anatomy (provides detailed information on eleven body systems and the special senses) *http://library.thinkquest.org/11965/*

The DNA Learning Center (provides in-depth information about genes for students and educators) *http://vector.cshl.org/*

Educational Hotlists at the Franklin Institute (provides extensive links and other resources on science subjects ranging from animals to wind energy) *http://sln.fi.edu/tfi/hotlists/hotlists.html*

ENC Web Links: Science (provides an extensive list of links to sites covering subject areas under earth and space science, physical science, life science, process skills, and the history of science) *http://www.enc.org/weblinks/science/*

ENC Web Links: Math topics (provides an extensive list of links to sites covering subject areas under topics such as advanced mathematics, algebra, geometry, data analysis and probability, applied mathematics, numbers and operations, measurement, and problem solving) *http://www.enc.org/weblinks/math/*

Encyclopaedia Britannica Discovering Dinosaurs Activity Guide *http://dinosaurs.eb.com/dinosaurs/study/*

The Exploratorium: The Museum of Science, Art, and Human Perception *http://www.exploratorium.edu/*

Where to Learn More

Where to Learn More

ExploreMath.com (provides highly interactive math activities for students and educators) *http://www.exploremath.com/*

ExploreScience.com (provides highly interactive science activities for students and educators) *http://www.explorescience.com/*

Imagine the Universe! (provides information about the universe for students aged 14 and up) *http://imagine.gsfc.nasa.gov/*

Mad Sci Network (highly searchable site provides extensive science information in addition to a search engine and a library to find science resources on the Internet; also allows students to submit questions to scientists) *http://www.madsci.org/*

The Math Forum (provides math-related information and resources for elementary through graduate-level students) *http://forum.swarthmore.edu/*

NASA Human Spaceflight: International Space Station (NASA homepage for the space station) *http://www.spaceflight.nasa.gov/station/*

NASA's Origins Program (provides up-to-the-minute information on the scientific quest to understand life and its place in the universe) *http://origins.jpl.nasa.gov/*

National Human Genome Research Institute (provides extensive information about the Human Genome Project) *http://www.nhgri.nih.gov:80/index.html*

New Scientist Online Magazine *http://www.newscientist.com/*

The Nine Planets (provides a multimedia tour of the history, mythology, and current scientific knowledge of each of the planets and moons in our solar system) *http://seds.lpl.arizona.edu/nineplanets/nineplanets/nineplanets.html*

The Particle Adventure (provides an interactive tour of quarks, neutrinos, antimatter, extra dimensions, dark matter, accelerators, and particle detectors) *http://particleadventure.org/*

PhysLink: Physics and astronomy online education and reference *http://physlink.com/*

Savage Earth Online (online version of the PBS series exploring earthquakes, volcanoes, tsunamis, and other seismic activity) *http://www.pbs.org/wnet/savageearth/*

Science at NASA (provides breaking information on astronomy, space science, earth science, and biological and physical sciences) *http://science.msfc.nasa.gov/*

Science Learning Network (provides Internet-guided science applications as well as many middle school science links) *http://www.sln.org/*

SciTech Daily Review (provides breaking science news and links to dozens of science and technology publications; also provides links to numerous "interesting" science sites) *http://www.scitechdaily.com/*

Space.com (space news, games, entertainment, and science fiction) *http://www.space.com/index.html*

SpaceDaily.com (provides latest news about space and space travel) *http://www.spacedaily.com/*

SpaceWeather.com (science news and information about the Sun-Earth environment) *http://www.spaceweather.com/*

The Why Files (exploration of the science behind the news; funded by the National Science Foundation) *http://whyfiles.org/*

Index

Italic type indicates volume numbers; **boldface** type indicates entries and their page numbers; (ill.) indicates illustrations.

A

Abacus   *1:* **1-2** 1 (ill.)
Abelson, Philip   *1:* 24
Abortion   *3:* 565
Abrasives   *1:* **2-4,** 3 (ill.)
Absolute dating   *4:* 616
Absolute zero   *3:* 595-596
Abyssal plains   *7:* 1411
Acceleration   *1:* **4-6**
Acetylsalicylic acid   *1:* **6-9,** 8 (ill.)
Acheson, Edward G.   *1:* 2
Acid rain   *1:* **9-14,** 10 (ill.), 12 (ill.), *6:* 1163, *8:* 1553
Acidifying agents   *1:* 66
Acids and bases   *1:* **14-16,** *8:* 1495
Acoustics   *1:* **17-23,** 17 (ill.), 20 (ill.)
Acquired immunodeficiency syndrome. *See* **AIDS (acquired immunodeficiency syndrome)**
Acrophobia   *8:* 1497
Actinides   *1:* **23-26,** 24 (ill.)
Acupressure   *1:* 121
Acupuncture   *1:* 121
Adams, John Couch   *7:* 1330
Adaptation   *1:* **26-32,** 29 (ill.), 30 (ill.)
Addiction   *1:* **32-37,** 35 (ill.), *3:* 478
Addison's disease   *5:* 801

Adena burial mounds   *7:* 1300
Adenosine triphosphate   *7:* 1258
ADHD   *2:* 237-238
Adhesives   *1:* **37-39,** 38 (ill.)
Adiabatic demagnetization   *3:* 597
ADP   *7:* 1258
Adrenal glands   *5:* 796 (ill.)
Adrenaline   *5:* 800
Aerobic respiration   *9:* 1673
Aerodynamics   *1:* **39-43,** 40 (ill.)
Aerosols   *1:* **43-49,** 43 (ill.)
Africa   *1:* **49-54,** 50 (ill.), 53 (ill.)
Afterburners   *6:* 1146
Agent Orange   *1:* **54-59,** 57 (ill.)
Aging and death   *1:* **59-62**
Agoraphobia   *8:* 1497
Agriculture   *1:* **62-65,** 63, 64 (ill.), *3:* 582-590, *5:* 902-903, *9:* 1743-744, *7:* 1433 (ill.)
Agrochemicals   *1:* **65-69,** 67 (ill.), 68 (ill.)
Agroecosystems   *2:* 302
AI. *See* **Artificial intelligence**
AIDS (acquired immunodeficiency syndrome)   *1:* **70-74,** 72 (ill.), *8:* 1583, *9:* 1737
Air flow   *1:* 40 (ill.)
Air masses and fronts   *1:* **80-82,** 80 (ill.)
Air pollution   *8:* 1552, 1558
Aircraft   *1:* **74-79,** 75 (ill.), 78 (ill.)
Airfoil   *1:* 41
Airplanes. *See* **Aircraft**
Airships   *1:* 75

Index

Al-jabr wa'l Muqabalah 1: 97
Al-Khwarizmi   1: 97
Alchemy   1: **82-85**
Alcohol (liquor)   1: 32, 85-87
Alcoholism   1: **85-88**
Alcohols   1: **88-91**, 89 (ill.)
Aldrin, Edwin   9: 1779
Ale   2: 354
Algae   1: **91-97**, 93 (ill.), 94 (ill.)
Algal blooms   1: 96
Algebra   1: **97-99**, 2: 333-334
Algorithms   1: 190
Alkali metals   1: **99-102**, 101 (ill.)
Alkaline earth metals   1: **102-106**, 104 (ill.)
Alleles   7: 1248
Allergic rhinitis   1: 106
Allergy   1: **106-110**, 108 (ill.)
Alloy   1: **110-111**
Alpha particles   2: 233, 8: 1620, 1632
Alps   5: 827, 7: 1301
Alternating current (AC)   4: 741
Alternation of generations   9: 1667
Alternative energy sources   1: **111-118**, 114 (ill.), 115 (ill.), 6: 1069
Alternative medicine   1: **118-122**
Altimeter   2: 266
Aluminum   1: 122-124, 125 (ill.)
Aluminum family   1: **122-126**, 125 (ill.)
Alzheimer, Alois   1: 127
Alzheimer's disease   1: **62, 126-130**, 128 (ill.)
Amazon basin   9: 1774
American Red Cross   2: 330
Ames test   2: 408
Amino acid   1: **130-131**
Aminoglycosides   1: 158
Ammonia   7: 1346
Ammonification   7: 1343
Amniocentesis   2: 322
Amoeba   1: **131-134**, 132 (ill.)
Ampere   3: 582, 4: 737
Amère, André   4: 737, 6: 1212
Ampere's law   4: 747
Amphibians   1: **134-137**, 136 (ill.)
Amphiboles   1: 191
Amphineura   7: 1289
Amplitude modulation   8: 1627
Amundsen, Roald   1: 152
Anabolism   7: 1255

Anaerobic respiration   9: 1676
Anatomy   1: **138-141**, 140 (ill.)
Anderson, Carl   1: 163, 4: 773
Andes Mountains   7: 1301, 9: 1775-1776
Andromeda galaxy   5: 939 (ill.)
Anemia   1: 8, 6: 1220
Aneroid barometer   2: 266
Anesthesia   1: **142-145**, 143 (ill.)
Angel Falls   9: 1774
Angiosperms   9: 1729
Animal behavior   2: 272
Animal hormones   6: 1053
Animal husbandry   7: 1433
Animals   1: **145-147**, 146 (ill.), 6: 1133-1134
Anorexia nervosa   4: 712
Antarctic Treaty   1: 153
Antarctica   1: **147-153**, 148 (ill.), 152 (ill.)
Antennas   1: **153-155**, 154 (ill.)
Anthrax   2: 287
Antibiotics   1: **155-159**, 157 (ill.)
Antibody and antigen   1: **159-162**, 2: 311
Anticyclones, cyclones and   3: 608-610
Antidiuretic hormone   5: 798
Antigens, antibodies and   1: 159-162
Antimatter   1: 163
Antimony   7: 1348
Antiparticles   1: **163-164**
Antiprotons   1: 163
Antipsychotic drugs   8: 1598
Antiseptics   1: **164-166**
Anurans   1: 136
Apennines   5: 827
Apes   8: 1572
Apgar Score   2: 322
Aphasia   9: 1798, 1799
Apollo 11   9: 1779, 1780 (ill.)
Apollo objects   1: 202
Appalachian Mountains   7: 1356
Appendicular skeleton   9: 1741
Aquaculture   1: **166-168**, 167 (ill.)
Arabian Peninsula.   *See* **Middle East**
Arabic numbers.   *See* **Hindu-Arabic number system**
Arachnids   1: **168-171**, 170 (ill.)
Arachnoid   2: 342
Arachnophobia   8: 1497
Ararat, Mount   1: 197

Index

Archaeoastronomy *1:* **171-173,** 172 (ill.)
Archaeology *1:* **173-177,** 175 (ill.), 176 (ill.), *7:* 1323-1327
Archaeology, oceanic. *See* **Nautical archaeology**
Archaeopteryx lithographica *2: 312*
Archimedes *2:* 360
Archimedes' Principle *2:* 360
Argon *7:* 1349, 1350
Ariel *10:* 1954
Aristotle *1:* 138, *2:* 291, *5:* 1012, *6:* 1169
Arithmetic *1:* 97, **177-181,** *3:* 534-536
Arkwright, Edmund *6:* 1098
Armstrong, Neil *9:* 1779
Arnold of Villanova *2:* 404
ARPANET *6:* 1124
Arrhenius, Svante *1:* 14, *8:* 1495
Arsenic *7:* 1348
Arthritis *1:* **181-183,** 182 (ill.)
Arthropods *1:* **183-186,** 184 (ill.)
Artificial blood *2:* 330
Artificial fibers *1:* **186-188,** 187 (ill.)
Artificial intelligence *1:* **188-190,** *2:* 244
Asbestos *1:* **191-194,** 192 (ill.), *6:* 1092
Ascorbic acid. *See* **Vitamin C**
Asexual reproduction *9:* 1664 (ill.), 1665
Asia *1:* **194-200,** 195 (ill.), 198 (ill.)
Aspirin. *See* **Acetylsalicylic acid**
Assembly language *3:* 551
Assembly line *7:* 1238
Astatine *6:* 1035
Asterisms *3:* 560
Asteroid belt *1:* 201
Asteroids *1:* **200-204,** 203 (ill.), *9:* 1764
Asthenosphere *8:* 1535, 1536
Asthma *1:* **204-207,** 206 (ill.), *9:* 1681
Aston, William *7:* 1240
Astronomia nova *3:* 425
Astronomy, infrared *6:* 1100-1103
Astronomy, ultraviolet *10:* 1943-1946
Astronomy, x-ray *10:* 2038-2041
Astrophysics *1:* **207-209,** 208 (ill.)
Atherosclerosis *3:* 484
Atmosphere observation *2:* **215-217,** 216 (ill.)

Atmosphere, composition and structure *2:* **211-215,** 214 (ill.)
Atmospheric circulation *2:* **218-221,** 220 (ill.)
Atmospheric optical effects *2:* **221-225,** 223 (ill.)
Atmospheric pressure *2:* **225,** 265, *8:* 1571
Atom *2:* **226-229,** 227 (ill.)
Atomic bomb *7:* 1364, 1381
Atomic clocks *10:* 1895-1896
Atomic mass *2:* 228, **229-232**
Atomic number *4:* 777
Atomic theory *2:* **232-236,** 234 (ill.)
ATP *7:* 1258
Attention-deficit hyperactivity disorder (ADHD) *2:* **237-238**
Audiocassettes. *See* **Magnetic recording/audiocassettes**
Auer metal *6:* 1165
Auroras *2:* 223, 223 (ill.)
Australia *2:* **238-242,** 239 (ill.), 241 (ill.)
Australopithecus afarensis *6:* 1056, 1057 (ill.)
Australopithecus africanus *6:* 1056
Autistic savants. *See* **Savants**
Autoimmune diseases *1:* 162
Automation *2:* **242-245,** 244 (ill.)
Automobiles *2:* **245-251,** 246 (ill.), 249 (ill.)
Autosomal dominant disorders *5:* 966
Auxins *6:* 1051
Avogadro, Amadeo *7:* 1282
Avogadro's number *7:* 1282
Axial skeleton *9:* 1740
Axioms *1:* 179
Axle *6:* 1207
Ayers Rock *2:* 240
AZT *1:* 73

B

B-2 Stealth Bomber *1:* 78 (ill.)
Babbage, Charles *3:* 547
Babbitt, Seward *9:* 1691
Bacitracin *1:* 158
Bacteria *2:* **253-260,** 255 (ill.), 256 (ill.), 259 (ill.)
Bacteriophages *10:* 1974

Index

Baekeland, Leo H.   *8:* 1565
Bakelite   *8:* 1565
Balard, Antoine   *6:* 1034
Baldwin, Frank Stephen   *2:* 371
Ballistics   *2:* **260-261**
Balloons   *1:* 75, *2:* ***261-265,*** 263 (ill.), 264 (ill.)
Bardeen, John   *10:* 1910
Barite   *6:* 1093
Barium   *1:* 105
Barnard, Christiaan   *6:* 1043, *10:* 1926
Barometer   *2:* **265-267,** 267 (ill.)
Barrier islands   *3:* 500
Bases, acids and 1: 14-16
Basophils   *2:* 329
Bats   *4:* 721
Battery   *2:* **268-270,** 268 (ill.)
Battle fatigue   *9:* 1826
Beaches, coasts and   *3:* 498-500
Becquerel, Henri   *8:* 1630
Bednorz, Georg   *10:* 1851
Behavior (human and animal), study of. *See* **Psychology**
Behavior   *2:* **270-273,** 271 (ill.), 272 (ill.)
Behaviorism (psychology)   *8:* 1595
Bell Burnell, Jocelyn   *7:* 1340
Bell, Alexander Graham   *10:* 1867 (ill.)
Benthic zone   *7:* 1415
Benz, Karl Friedrich   *2:* 246 (ill.)
Berger, Hans   *9:* 1745
Beriberi   *6:* 1219, *10:* 1982
Bernoulli's principle   *1:* 40, 42, *5:* 884
Beryllium   *1:* 103
Berzelius, Jöns Jakob   *2:* 230
Bessemer converter   *7:* 1445, *10:* 1916
Bessemer, Henry   *10:* 1916
Beta carotene   *10:* 1984
Beta particles   *8:* 1632
Bichat, Xavier   *1:* 141
Big bang theory   *2:* **273-276,** 274 (ill.), *4:* 780
Bigelow, Julian   *3:* 606
Binary number system   *7:* 1397
Binary stars   *2:* **276-278,** 278 (ill.)
Binomial nomenclature   *2:* 337
Biochemistry   *2:* **279-280**
Biodegradable   *2:* **280-281**
Biodiversity   *2:* **281-283,** 282 (ill.)
Bioenergy   *1:* 117, *2:* **284-287,** 284 (ill.)

Bioenergy fuels   *2:* 286
Biofeedback   *1:* 119
Biological warfare   *2:* **287-290**
Biological Weapons Convention Treaty   *2:* 290
Biology   *2:* **290-293,** *7:* 1283-1285
Bioluminescence   *6:* 1198
Biomass energy.   *See* **Bioenergy**
Biomes   *2:* **293-302,** 295 (ill.), 297 (ill.), 301 (ill.)
Biophysics   *2:* **302-304**
Bioremediation   *7:* 1423
Biosphere 2 Project   *2:* 307-309
Biospheres   *2:* **304-309,** 306 (ill.)
Biot, Jean-Baptiste   *7:* 1262
Biotechnology   *2:* **309-312,** 311 (ill.)
Bipolar disorder   *4:* 633
Birds   *2:* **312-315,** 314 (ill.)
Birth   *2:* **315-319,** 317 (ill.), 318 (ill.)
Birth control.   *See* **Contraception**
Birth defects   *2:* **319-322,** 321 (ill.)
Bismuth   *7:* 1349
Bjerknes, Jacob   *1:* 80, *10:* 2022
Bjerknes, Vilhelm   *1:* 80, *10:* 2022
Black Death   *8:* 1520
Black dwarf   *10:* 2028
Black holes   *2:* **322-326,** 325 (ill.), *9:* 1654
Blanc, Mont   *5:* 827
Bleuler, Eugen   *9:* 1718
Blood   *2:* **326-330,** 328 (ill.), 330, *3:* 483
Blood banks   *2:* 330
Blood pressure   *3:* 483
Blood supply   *2:* **330-333**
Blood vessels   *3:* 482
Blue stars   *9:* 1802
Bode, Johann   *1:* 201
Bode's Law   *1:* 201
Bogs   *10:* 2025
Bohr, Niels   *2:* 235
Bones.   *See* **Skeletal system**
Bones, study of diseases of or injuries to. *See* **Orthopedics**
Boole, George   *2:* 333
Boolean algebra   *2:* **333-334**
Bopp, Thomas   *3:* 529
Borax   *1:* 126, *6:* 1094
Boreal coniferous forests   *2:* 294
Bores, Leo   *8:* 1617
Boron   *1:* 124-126

Index

Boron compounds *6:* 1094
Bort, Léon Teisserenc de *10:* 2021
Bosons *10:* 1831
Botany *2:* **334-337,** 336 (ill.)
Botulism *2:* 258, 288
Boundary layer effects *5:* 885
Bovine growth hormone *7:* 1434
Boyle, Robert *4:* 780
Boyle's law *5:* 960
Braham, R. R. *10:* 2022
Brahe, Tycho *3:* 574
Brain *2:* **337-351,** 339 (ill.), 341 (ill.)
Brain disorders *2:* 345
Brass *10:* 1920
Brattain, Walter *10:* 1910
Breathing *9:* 1680
Brewing *2:* **352-354,** 352 (ill.)
Bridges *2:* **354-358,** 357 (ill.)
Bright nebulae *7:* 1328
British system of measurement *10:* 1948
Bromine *6:* 1034
Bronchitis *9:* 1681
Bronchodilators *1:* 205
Brønsted, J. N. *1:* 15
Brønsted, J. N. *1:* 15
Bronze *2:* 401
Bronze Age *6:* 1036
Brown algae *1:* 95
Brown dwarf *2:* **358-359**
Brucellosis *2:* 288
Bryan, Kirk *8:* 1457
Bubonic plague *8:* 1518
Buckminsterfullerene *2:* 398, 399 (ill.)
Bugs. *See* **Insects**
Bulimia *4:* 1714-1716
Buoyancy *1:* 74, *2:* **360-361,** 360 (ill.)
Burial mounds *7:* 1298
Burns *2:* **361-364,** 362 (ill.)
Bushnell, David *10:* 1834
Butterflies *2:* **364-367,** 364 (ill.)
Byers, Horace *10:* 2022

C

C-12 *2:* 231
C-14 *1:* 176, *4:* 617
Cable television *10:* 1877
Cactus *4:* 635 (ill.)
CAD/CAM *2:* **369-370,** 369 (ill.)

Caffeine *1:* 34
Caisson *2:* 356
Calcite *3:* 422
Calcium *1:* 104 (ill.), 105
Calcium carbonate *1:* 104 (ill.)
Calculators *2:* **370-371,** 370 (ill.)
Calculus *2:* **371-372**
Calderas *6:* 1161
Calendars *2:* **372-375,** 374 (ill.)
Callisto *6:* 1148, 1149
Calories *2:* **375-376,** *6:* 1045
Calving (icebergs) *6:* 1078, 1079 (ill.)
Cambium *10:* 1927
Cambrian period *8:* 1461
Cameroon, Mount *1:* 51
Canadian Shield *7:* 1355
Canals *2:* **376-379,** 378 (ill.)
Cancer *2:* **379-382,** 379 (ill.), 381 (ill.), *10:* 1935
Canines *2:* **382-387,** 383 (ill.), 385 (ill.)
Cannabis sativa *6:* 1224, 1226 (ill.)
Cannon, W. B. *8:* 1516
Capacitor *4:* 749
Carbohydrates *2:* **387-389,** *7:* 1400
Carbon *2:* 396
Carbon compounds, study of. *See* **Organic chemistry**
Carbon cycle *2:* **389-393,** 391 (ill.)
Carbon dioxide *2:* **393-395,** 394 (ill.)
Carbon family *2:* **395-403,** 396 (ill.), 397 (ill.), 399 (ill.)
Carbon monoxide *2:* **403-406**
Carbon-12 *2:* 231
Carbon-14 *4:* 617
Carboniferous period *8:* 1462
Carborundum *1:* 2
Carcinogens *2:* **406-408**
Carcinomas *2:* 381
Cardano, Girolamo *8:* 1576
Cardiac muscle *7:* 1312
Cardiovascular system *3:* 480
Caries *4:* 628
Carlson, Chester *8:* 1502, 1501 (ill.)
Carnot, Nicholas *6:* 1118
Carnot, Sadi *10:* 1885
Carothers, Wallace *1:* 186
Carpal tunnel syndrome *2:* **408-410**
Cartography *2:* **410-412,** 411 (ill.)
Cascade Mountains *7:* 1358
Caspian Sea *5:* 823, 824

Index

Cassini division *9:* 1711
Cassini, Giovanni Domenico *9:* 1711
Cassini orbiter *9:* 1712
CAT scans *2:* 304, *8:* 1640
Catabolism *7:* 1255
Catalysts and catalysis *3:* **413-415**
Catastrophism *3:* **415**
Cathode *3:* **415-416**
Cathode-ray tube *3:* **417-420,** 418 (ill.)
Cats. *See* **Felines**
Caucasus Mountains *5:* 823
Cavendish, Henry *6:* 1069, *7:* 1345
Caves *3:* **420-423,** 422 (ill.)
Cavities (dental) *4:* 628
Cayley, George *1:* 77
CDC *6:* 1180
CDs. *See* **Compact disc**
Celestial mechanics *3:* **423-428,** 427 (ill.)
Cell wall (plants) *3:* 436
Cells *3:* **428-436,** 432 (ill.), 435 (ill.)
Cells, electrochemical *3:* **436-439**
Cellular metabolism *7:* 1258
Cellular/digital technology *3:* **439-441**
Cellulose *2:* 389, *3:* **442-445,** 442 (ill.)
Celsius temperature scale *10:* 1882
Celsius, Anders *10:* 1882
Cenozoic era *5:* 990, *8:* 1462
Center for Disease Control (CDC) *6:* 1180
Central Asia *1:* 198
Central Dogma *7:* 1283
Central Lowlands (North America) *7:* 1356
Centrifuge *3:* **445-446,** 446 (ill.)
Cephalopoda *7:* 1289
Cephalosporin *1:* 158
Cepheid variables *10:* 1964
Ceramic *3:* **447-448**
Cerebellum *2:* 345
Cerebral cortex *2:* 343
Cerebrum *2:* 343
Čerenkov effect *6:* 1189
Cerium *6:* 1163
Cesium *1:* 102
Cetaceans *3:* **448-451,** 450 (ill.), *4:* 681 (ill.), *7:* 1416 (ill.)
CFCs *6:* 1032, *7:* 1453-1454, *8:* 1555,
Chadwick, James *2:* 235, *7:* 1338
Chain, Ernst *1:* 157

Chamberlain, Owen *1:* 163
Chancroid *9:* 1735, 1736
Chandra X-ray Observatory *10:* 2040
Chandrasekhar, Subrahmanyan *10:* 1854
Chandrasekhar's limit *10:* 1854
Chao Phraya River *1:* 200
Chaos theory *3:* **451-453**
Chaparral *2:* 296
Chappe, Claude *10:* 1864
Chappe, Ignace *10:* 1864
Charles's law *5:* 961
Charon *8:* 1541, 1541 (ill.), 1542
Chassis *2:* 250
Cheetahs *5:* 861
Chemical bond *3:* **453-457**
Chemical compounds *3:* 541-546
Chemical elements *4:* 774-781
Chemical equations *5 :* *815-817*
Chemical equilibrium *5:* 817-820
Chemical warfare *3:* **457-463,** 459 (ill.), 461 (ill.)*6:* 1032
Chemiluminescence *6:* 1198
Chemistry *3:* **463-469,** 465 (ill.) ,467 (ill.), *8:* 1603
Chemoreceptors *8:* 1484
Chemosynthesis *7:* 1418
Chemotherapy *2:* 382
Chichén Itzá *1:* 173
Chicxulub *1:* 202
Childbed fever *1:* 164
Chimpanzees *8:* 1572
Chiropractic *1:* 120
Chladni, Ernst *1:* 17
Chlamydia *9:* 1735, 1736
Chlorination *6:* 1033
Chlorine *6:* 1032
Chlorofluorocarbons. *See* **CFCs**
Chloroform *1:* 142, 143, 143 (ill.)
Chlorophyll *1:* 103
Chlorophyta *1:* 94
Chloroplasts *3:* 436, *8:* 1506 (ill.)
Chlorpromazine *10:* 1906
Cholesterol *3:* **469-471,** 471 (ill.), *6:* 1042
Chorionic villus sampling *2:* 322, *4:* 790
Chromatic aberration *10:* 1871
Chromatography *8:* 1604
Chromosomes *3:* **472-476,** 472 (ill.), 475(ill.)

Index

Chromosphere *10:* 1846
Chrysalis *2:* 366, *7:* 1261 (ill.)
Chrysophyta *1:* 93
Chu, Paul Ching-Wu *10:* 1851
Cigarette smoke *3:* **476-478,** 477 (ill.)
Cigarettes, addiction to *1:* 34
Ciliophora *8:* 1592
Circle *3:* **478-480,** 479 (ill.)
Circular acceleration *1:* 5
Circular accelerators *8:* 1479
Circulatory system *3:* **480-484,** 482 (ill.)
Classical conditioning *9:* 1657
Clausius, Rudolf *10:* 1885
Claustrophobia *8:* 1497
Climax community *10:* 1839
Clones and cloning *3:* **484-490,** 486 (ill.), 489 (ill.)
Clostridium botulinum *2:* 258
Clostridium tetani *2:* 258
Clouds *3:* **490-492,** 491 (ill.)
Coal *3:* **492-498,** 496 (ill.)
Coast and beach *3:* **498-500.** 500 (ill.)
Coastal Plain (North America) *7:* 1356
Cobalt-60 *7:* 1373
COBE (Cosmic Background Explorer) *2:* 276
COBOL *3:* 551
Cocaine *1:* 34, *3:* **501-505,** 503 (ill.)
Cockroaches *3:* **505-508,** 507 (ill.)
Coelacanth *3:* **508-511,** 510 (ill.)
Cognition *3:* **511-515,** 513 (ill.), 514 (ill.)
Cold fronts *1:* 81, 81 (ill.)
Cold fusion *7:* 1371
Cold-deciduous forests *5:* 909
Collins, Francis *6:* 1064
Collins, Michael *9:* 1779
Colloids *3:* **515-517,** 517 (ill.)
Color *3:* **518-522,** 521 (ill.)
Color blindness *5:* 971
Colorant *4:* 686
Colt, Samuel *7:* 1237
Columbus, Christopher *1:* 63
Coma *2:* 345
Combined gas law *5:* 960
Combustion *3:* **522-527,** 524 (ill.), *7:* 1441
Comet Hale-Bopp *3:* 529
Comet Shoemaker-Levy 9 *6:* 1151
Comet, Halley's *3:* 528

Comets *3:* **527-531,** 529 (ill.), *6:* 1151, *9:* 1765
Common cold *10:* 1978
Compact disc *3:* **531-533,** 532 (ill.)
Comparative genomics *6:* 1067
Complex numbers *3:* **534-536,** 534 (ill.), *6:* 1082
Composite materials *3:* **536-539**
Composting *3:* **539-541,** 539 (ill.)
Compound, chemical *3:* **541-546,** 543 (ill.)
Compton Gamma Ray Observatory *5:* 949
Compulsion *7:* 1405
Computer Aided Design and Manufacturing. *See* **CAD/CAM**
Computer languages *1:* 189, *3:* 551
Computer software *3:* **549-554,** 553 (ill.)
Computer, analog *3:* **546-547**
Computer, digital *3:* **547-549,** 548 (ill.)
Computerized axial tomography. *See* **CAT scans**
Concave lenses *6:* 1185
Conditioning *9:* 1657
Condom *3:* 563
Conduction *6:* 1044
Conductivity, electrical. *See* **Electrical conductivity**
Conservation laws *3:* **554-558.** 557 (ill.)
Conservation of electric charge *3:* 556
Conservation of momentum *7:* 1290
Conservation of parity *3:* 558
Constellations *3:* **558-560,** 559 (ill.)
Contact lines *5:* 987
Continental Divide *7:* 1357
Continental drift *8:* 1534
Continental margin *3:* **560-562**
Continental rise *3:* 562
Continental shelf *2:* 300
Continental slope *3:* 561
Contraception *3:* **562-566,** 564 (ill.)
Convection *6:* 1044
Convention on International Trade in Endangered Species *5:* 795
Convex lenses *6:* 1185
Cooke, William Fothergill *10:* 1865
Coordination compounds *3:* 544
Copernican system *3:* 574
Copper *10:* 1919-1921, 1920 (ill.)

Index

Coral *3:* **566-569,** 567 (ill.), 568 (ill.)
Coral reefs *2:* 301
Core *4:* 711
Coriolis effect *2:* 219, *10:* 2029
Corona *10:* 1846
Coronary artery disease *6:* 1042
Coronas *2:* 225
Correlation *3:* **569-571**
Corson, D. R. *6:* 1035
Corti, Alfonso Giacomo Gaspare *4:* 695
Corticosteroids *1:* 206
Corundum *6:* 1094
Cosmetic plastic surgery *8:* 1530
Cosmic Background Explorer (COBE) *2:* 276
Cosmic dust *6:* 1130
Cosmic microwave background *2:* 275, *8:* 1637
Cosmic rays *3:* **571-573,** 573 (ill.)
Cosmology *1:* 171, *3:* **574-577**
Cotton *3:* **577-579,** 578 (ill.)
Coulomb *3:* **579-582**
Coulomb, Charles *3:* 579, *6:* 1212
Coulomb's law *4:* 744
Courtois, Bernard *6:* 1035
Courtship behaviors *2:* 273
Covalent bonding *3:* 455
Cowan, Clyde *10:* 1833
Coxwell, Henry Tracey *2:* 263
Coyotes *2:* 385
Craniotomy *8:* 1528
Creationism *3:* 577
Crick, Francis *3:* 473, *4:* 786, *5:* 973, 980 (ill.), 982, *7:* 1389
Cro-Magnon man *6:* 1059
Crop rotation *3:* 589
Crops *3:* **582-590,** 583 (ill.), 589 (ill.)
Crude oil *8:* 1492
Crust *4:* 709
Crustaceans *3:* **590-593,** 592 (ill.)
Cryobiology *3:* **593-595**
Cryogenics *3:* **595-601,** 597 (ill.)
Crystal *3:* **601-604,** 602 (ill.), 603 (ill.)
Curie, Marie *7:* 1450
Current electricity *4:* 742
Currents, ocean *3:* **604-605**
Cybernetics *3:* **605-608,** 607 (ill.)
Cyclamate *3:* **608**
Cyclone and anticyclone *3:* **608-610,** 609 (ill.)

Cyclotron *1:* 163, *8:* 1479, 1480 (ill.)
Cystic fibrosis *2:* 320
Cytokinin *6:* 1052
Cytoskeleton *3:* 434

D

Da Vinci, Leonardo *2:* 291, *4:* 691, *10:* 2020
Daddy longlegs *1:* 171
Dalton, John *2:* 226, 229, *2:* 232
Dalton's theory *2:* 232
Dam *4:* **611-613,** 612 (ill.)
Damselfly *1:* 184 (ill.)
Danube River *5:* 824
Dark matter *4:* **613-616,** 615 (ill.)
Dark nebulae *6:* 1131, *7:* 1330
Dart, Raymond *6:* 1056
Darwin, Charles *1:* 29, *6:* 1051, *8:* 1510
Dating techniques *4:* **616-619,** 618 (ill.)
Davy, Humphry *1:* 142, chlorine *6:* 1032, 1087
DDT (dichlorodiphenyltrichloroethane) *1:* 69, *4:* **619-622,** 620 (ill.)
De Bort, Léon Philippe Teisserenc *2:* 263
De Candolle, Augustin Pyrame *8:* 1509
De curatorum chirurgia *8:* 1528
De Forest, Lee *10:* 1961
De materia medica *5:* 877
De Soto, Hernando *7:* 1299
Dead Sea *1:* 196
Death *1:* 59-62
Decay *7:* 1442
Decimal system *1:* 178
Decomposition *2:* 392, *9:* 1648
Deimos *6:* 1229
Dementia *4:* **622-624,** 623 (ill.)
Democritus *2:* 226, 232
Dendrochronology. *See* **Tree-ring dating**
Denitrification *7:* 1343
Density *4:* **624-626,** 625 (ill.)
Dentistry *4:* **626-630,** 628 (ill.), 629 (ill.)
Depression *4:* **630-634,** 632 (ill.)
Depth perception *8:* 1483 (ill.), 1484

Index

Dermis *6:* 1111
Desalination *7:* 1439, *10:* 2012
Descartes, René *6:* 1184
The Descent of Man *6:* 1055
Desert *2:* 296, *4:* **634-638,** 635 (ill.), 636 (ill.)
Detergents, soaps and *9:* 1756-1758
Devonian period *8:* 1461
Dew point *3:* 490
Dexedrine *2:* 238
Diabetes mellitus *4:* **638-640**
Diagnosis *4:* **640-644,** 643 (ill.)
Dialysis *4:* **644-646,** *7:* 1439
Diamond *2:* 396 (ill.), 397
Diencephalon *2:* 342
Diesel engine *4:* **646-647,** 647 (ill.)
Diesel, Rudolf *4:* 646, *10:* 1835
Differential calculus *2:* 372
Diffraction *4:* **648-651,** 648 (ill.)
Diffraction gratings *4:* 650
Diffusion *4:* **651-653,** 652 (ill.)
Digestion *7:* 1255
Digestive system *4:* **653-658,** 657 (ill.)
Digital audio tape *6:* 1211
Digital technology. *See* **Cellular/digital technology**
Dingoes *2:* 385, 385 (ill.)
Dinosaurs *4:* **658-665,** 660 (ill.), 663 (ill.), 664 (ill.)
Diodes *4:* **665-666,** *6:* 1176-1179
Dioscorides *5:* 878
Dioxin *4:* **667-669**
Dirac, Paul *1:* 163, *4:* 772
Dirac's hypothesis *1:* 163
Direct current (DC) *4:* 741
Dirigible *1:* 75
Disaccharides *2:* 388
Disassociation *7:* 1305
Disease *4:* **669-675,** 670 (ill.), 673 (ill.), *8:* 1518
Dissection *10:* 1989
Distillation *4:* **675-677,** 676 (ill.)
DNA *1:* 61, *2:* 310, *3:* 434, 473-474, *5:* 972-975, 980 (ill.), 981-984, *7:* 1389-1390
  forensic science *5:* 900
  human genome project *6:* 1060-1068
  mutation *7:* 1314-1316
Döbereiner, Johann Wolfgang *8:* 1486
Dogs. *See* **Canines**

Dollard, John *10:* 1871
Dolly (clone) *3:* 486
Dolphins *3:* 448, 449 (ill.)
Domagk, Gerhard *1:* 156
Domain names (computers) *6:* 1127
Dopamine *9:* 1720
Doppler effect *4:* **677-680,** 679 (ill.), *9:* 1654
Doppler radar *2:* 220 (ill.), *10:* 2023
Doppler, Christian Johann *9:* 1654
Down syndrome *2:* 319
Down, John Langdon Haydon *9:* 1713
Drake, Edwin L. *7:* 1419
Drebbel, Cornelius *10:* 1834
Drew, Richard *1:* 39
Drift nets *4:* **680-682,** 681 (ill.)
Drinker, Philip *8:* 1548
Drought *4:* **682-684,** 683 (ill.)
Dry cell (battery) *2:* 269
Dry ice *2:* 395
Drying (food preservation) *5:* 890
Dubois, Marie-Eugene *6:* 1058
Duodenum *4:* 655
Dura mater *2:* 342
Dust Bowl *4:* 682
Dust devils *10:* 1902
Dust mites *1:* 107, 108 (ill.)
DVD technology *4:* **684-686**
Dyes and pigments *4:* **686-690,** 688 (ill.)
Dynamite *5:* 845
Dysarthria *9:* 1798
Dyslexia *4:* **690-691,** 690 (ill.)
Dysphonia *9:* 1798
Dysprosium *6:* 1163

E

$E = mc^2$ *7:* 1363, 1366, *9:* 1662
Ear *4:* **693-698,** 696 (ill.)
Earth (planet) *4:* **698-702,** 699 (ill.)
Earth science *4:* **707-708**
Earth Summit *5:* 796
Earth's interior *4:* **708-711,** 710 (ill.)
Earthquake *4:* **702-707,** 705 (ill.), 706 (ill.)
Eating disorders *4:* **711-717,** 713 (ill.)
Ebola virus *4:* **717-720,** 719 (ill.)
Echolocation *4:* **720-722**
Eclipse *4:* **723-725,** 723 (ill.)

Index

Ecological pyramid *5:* 894 (ill.), 896
Ecological system. *See* **Ecosystem**
Ecologists *4:* 728
Ecology *4:* **725-728**
Ecosystem *4:* **728-730,** 729 (ill.)
Edison, Thomas Alva *6:* 1088
EEG (electroencephalogram) *2:* 348, *9:* 1746
Eijkman, Christian *10:* 1981
Einstein, Albert *4:* 691, *7:* 1428, *9:* 1659 (ill.)
　photoelectric effect *6:* 1188, *8:* 1504
　space-time continuum *9:* 1777
　theory of relativity *9:* 1659-1664
Einthoven, William *4:* 751
EKG (electrocardiogram) *4:* 751-755
El Niño *4:* **782-785,** 784 (ill.)
Elasticity *4:* **730-731**
Elbert, Mount *7:* 1357
Elbrus, Mount *5:* 823
Electric arc *4:* **734-737,** 735 (ill.)
Electric charge *4:* 743
Electric circuits *4:* 739, 740 (ill.)
Electric current *4:* 731, 734, **737-741,** 740 (ill.), 746, 748, 761, 767, 771, 773
Electric fields *4:* 743, 759
Electric motor *4:* **747-750,** 747 (ill.)
Electrical conductivity *4:* **731-734,** 735
Electrical force *3:* 579, 581-582, *4:* 744
Electrical resistance *4:* 732, 738, 746
Electricity *4:* **741-747,** 745 (ill.)
Electrocardiogram *4:* **751-755,** 753 (ill.), 754 (ill.)
Electrochemical cells *3:* 416, 436-439
Electrodialysis *4:* 646
Electroluminescence *6:* 1198
Electrolysis *4:* **755-758**
Electrolyte *4:* 755
Electrolytic cell *3:* 438
Electromagnet *6:* 1215
Electromagnetic field *4:* **758-760**
Electromagnetic induction *4:* **760-763,** 762 (ill.)
Electromagnetic radiation *8:* 1619
Electromagnetic spectrum *4:* **763-765,** *4:* 768, *6:* 1100, 1185, *8:* 1633, *9:* 1795
Electromagnetic waves *7:* 1268
Electromagnetism *4:* **766-768,** 766 (ill.)
Electron *4:* **768-773**
Electron gun *3:* 417
Electronegativity *3:* 455
Electronics *4:* **773-774,** 773 (ill.)
Electrons *4:* **768-773,** *10:* 1832, 1833
Electroplating *4:* 758
Element, chemical *4:* **774-781,** 778 (ill.)
Elementary algebra *1:* 98
Elements *4:* 775, 777, *8:* 1490, *10:* 1913
Embryo and embryonic development *4:* **785-791,** 788 (ill.)
Embryology *4:* 786
Embryonic transfer *4:* 790-791
Emphysema *9:* 1681
Encke division *9:* 1711
Encke, Johann *9:* 1711
Endangered species *5:* **793-796,** 795 (ill.)
Endangered Species Act *5:* 795
Endocrine system *5:* **796-801,** 799 (ill.)
Endoplasmic reticulum *3:* 433
Energy *5:* **801-805**
Energy and mass *9:* 1662
Energy conservation *1:* 117
Energy, alternative sources of *1:* 111-118, *6:* 1069
Engels, Friedrich *6:* 1097
Engineering *5:* **805-807,** 806 (ill.)
Engines *2:* 246, *6:* 1117, 1143, *9:* 1817, *10:* 1835
English units of measurement. *See* **British system of measurement**
ENIAC *3:* 551
Entropy *10:* 1886
Environment
　air pollution *8:* 1552, 1553
　effect of aerosols on *1:* 47, 48
　effect of carbon dioxide on *8:* 1554
　effect of use of fossil fuels on *2:* 285, *7:* 1454
　impact of aquaculture on *1:* 168
　industrial chemicals *8:* 1557
　ozone depletion *8:* 1555
　poisons and toxins *8:* 1546
　tropical deforestation *9:* 1744
　water pollution *8:* 1556
Environmental ethics *5:* **807-811,** 809 (ill.), 810 (ill.)

Enzyme *5:* **812-815,** 812 (ill.), 814 (ill.)
Eosinophils *2:* 329
Epidemics *4:* 671
Epidermis *2:* 362, *6:* 1110
Epilepsy *2:* 347-349
Equation, chemical *5:* **815-817**
Equilibrium, chemical *5:* **817-820**
Equinox *9:* 1728
Erasistratus *1:* 138
Erbium *6:* 1163
Erosion *3:* 498, *5:* **820-823,** 821 (ill.), *9:* 1762
Erythroblastosis fetalis *9:* 1685
Erythrocytes *2:* 327
Escherichia coli *2:* 258
Esophagitis *4:* 656
Estrogen *5:* 801, *8:* 1599, 1600
Estuaries *2:* 300
Ethanol *1:* 89-91
Ether *1:* 142, 143
Ethics *3:* 489, *5:* 807-811
Ethylene glycol *1:* 91
Euglenoids *1:* 92
Euglenophyta *1:* 92
Eukaryotes *3:* 429, 432-435
Europa *6:* 1148, 1149
Europe *5:* **823-828,** 825 (ill.), 827 (ill.)
Europium *6:* 1163
Eutrophication *1:* 96, *5:* **828-831,** 830 (ill.)
Evans, Oliver *7:* 1237, *9:* 1820
Evaporation *5:* **831-832**
Everest, Mount *1:* 194
Evergreen broadleaf forests *5:* 909
Evergreen tropical rain forest *2:* 298
Evolution *1:* 26, 51, *5:* **832-839**
Excretory system *5:* **839-842**
Exhaust system *2:* 247
Exoplanets. *See* **Extrasolar planets**
Exosphere *2:* 214
Expansion, thermal *5:* 842-843, *10:* 1883-1884
Expert systems *1:* 188
Explosives *5:* **843-847**
Extrasolar planets *5:* **847-848,** 846 (ill.)
Extreme Ultraviolet Explorer *6:* 1123
Exxon *Valdez* 7: *1424, 1425 (ill.)*
Eye *5:* **848-853,** 851 (ill.)
Eye surgery *8:* 1615-1618

F

Fahrenheit temperature scale *10:* 1882
Fahrenheit, Gabriel Daniel *10:* 1882
Far East *1:* 199
Faraday, Michael *4:* 761, 767, *6:* 1212
Farming. *See* **Agriculture**
Farnsworth, Philo *10:* 1875
Farsightedness *5:* 851
Father of
 acoustics *1:* 17
 American psychiatry *9:* 1713
 genetics *5:* 982
 heavier-than-air craft *1:* 77
 lunar topography *7:* 1296
 medicine *2:* 348
 modern chemistry *3:* 465
 modern dentistry *4:* 627
 modern evolutionary theory *5:* 833
 modern plastic surgery *8:* 1529
 rigid airships *1:* 75
 thermochemistry *3:* 525
Fats *6:* 1191
Fauchard, Pierre *4:* 627
Fault *5:* **855,** 856 (ill.)
Fault lines *5:* 987
Fear, abnormal or irrational. *See* **Phobias**
Feldspar *6:* 1094
Felines *5:* **855-864,** 861, 862 (ill.)
Fermat, Pierre de *7:* 1393, *8:* 1576
Fermat's last theorem *7:* 1393
Fermentation *5:* **864-867,** *10:* 2043
Fermi, Enrico *7:* 1365
Ferrell, William *2:* 218
Fertilization *5:* **867-870,** 868 (ill.)
Fertilizers *1:* 66
Fetal alcohol syndrome *1:* 87
Fiber optics *5:* **870-872,** 871 (ill.)
Fillings (dental) *4:* 628
Filovirus *4:* 717
Filtration *5:* **872-875**
Fingerprinting *5:* 900
Fire algae *1:* 94
First law of motion *6:* 1170
First law of planetary motion *7:* 1426
First law of thermodynamics *10:* 1885
Fish *5:* **875-878,** 876 (ill.)
Fish farming *1:* 166
Fishes, age of *8:* 1461
FitzGerald, George Francis *9:* 1660

Index

Index

Flash lock *6:* 1193
Fleas *8:* 1474, 1474 (ill.)
Fleischmann, Martin *7:* 1371
Fleming, Alexander *1:* 156
Fleming, John Ambrose *10:* 1961
Florey, Howard *1:* 157
Flower *5:* **878-862,** 881 (ill.)
Flu. *See* **Influenza**
Fluid dynamics *5:* **882-886**
Flukes *8:* 1473
Fluorescence *6:* 1197
Fluorescent light *5:* **886-888,** 888 (ill.)
Fluoridation *5:* **889-890**
Fluoride *5:* 889
Fluorine *6:* 1031-1032
Fluorspar *6:* 1095
Fly shuttle *6:* 1097
Fold lines *5:* 987
Food irradiation *5:* 893
Food preservation *5:* **890-894**
Food pyramid *7:* 1402, 1402 (ill.)
Food web and food chain *5:* **894-898,** 896 (ill.)
Ford, Henry *2:* 249 (ill.), *7:* 1237-1238
Forensic science *5:* **898-901,** 899 (ill.), *6:* 1067
Forestry *5:* **901-907,** 905 (ill.), 906 (ill.)
Forests *2:* 294-295, *5:* **907-914,** 909 (ill.), 910 (ill.), 913 (ill.)
Formula, chemical *5:* **914-917**
FORTRAN *3:* 551
Fossil and fossilization *5:* **917-921,** 919 (ill.), 920 (ill.), *6:* 1055, *7:* 1326 (ill.), *8:* 1458
Fossil fuels *1:* 112, *2:* 284, 392, *7:* 1319
Fossils, study of. *See* **Paleontology**
Foxes *2:* 384
Fractals *5:* **921-923,** 922 (ill.)
Fractions, common *5:* **923-924**
Fracture zones *7:* 1410
Francium *1:* 102
Free radicals *1:* 61
Freezing point *3:* 490
Frequency *4:* 763, *5:* **925-926**
Frequency modulation *8:* 1628
Freshwater biomes *2:* 298
Freud, Sigmund *8:* 1593, 1594
Friction *5:* **926-927**
Frisch, Otto *7:* 1362

Fronts *1:* 80-82
Fry, Arthur *1:* 39
Fujita Tornado Scale *10:* 1902
Fujita, T. Theodore *10:* 1902
Fuller, R. Buckminster *2:* 398
Fulton, Robert *10:* 1835
Functions (mathematics) *5:* **927-930,** *8:* 1485
Functional groups *7:* 1430
Fungi *5:* **930-934,** 932 (ill.)
Fungicides *1:* 67
Funk, Casimir *10:* 1982
Fyodorov, Svyatoslav N. *8:* 1617

G

Gabor, Dennis *6:* 1049
Gadolinium *6:* 1163
Gagarin, Yury *9:* 1778
Gaia hypothesis *5:* **935-940**
Galactic clusters *9:* 1808
Galaxies, active *5:* 944
Galaxies *5:* **941-945,** 941 (ill.), 943 (ill.), *9:* 1806-1808
Galen, Claudius *1:* 139
Galileo Galilei *1:* 4, *5:* *1012*, *6: 1149, 1170, 1184,* *7: 1296,* 10: *1869*
Galileo probe *6:* 1149
Gall bladder *3:* 469, *4:* 653, 655
Galle, Johann *7:* 1330
Gallium *1:* 126
Gallo, Robert *10:* 1978
Gallstones *3:* 469
Galvani, Luigi *2:* 304, *4:* 751
Gambling *1:* 36
Game theory *5:* **945-949**
Gamma rays *4:* 765, *5:* **949-951,** *8:* 1632
Gamma-ray burst *5:* **952-955,** 952 (ill.), 954 (ill.)
Ganges Plain *1:* 197
Ganymede *6:* 1148, 1149
Garbage. *See* **Waste management**
Gardening. *See* **Horticulture**
Gas, natural *7:* 1319-1321
Gases, electrical conductivity in *4:* 735
Gases, liquefaction of *5:* **955-958**
Gases, properties of *5:* **959-962,** 959 (ill.)
Gasohol *1:* 91

Index

Gastropoda   *7:* 1288
Gauss, Carl Friedrich   *6:* 1212
Gay-Lussac, Joseph Louis   *2:* 262
Gay-Lussac's law   *5:* 962
Geiger counter   *8:* 1625
Gell-Mann, Murray   *10:* 1829
Generators   *5:* **962-966,** 964 (ill.)
Genes   *7:* 1248
Genes, mapping.   *See* **Human Genome Project**
Genetic disorders   *5:* **966-973,** 968 (ill.), 968 (ill.)
Genetic engineering   *2:* 310, *5:* **973-980,** 976 (ill.), 979 (ill.)
Genetic fingerprinting   *5:* 900
Genetics   *5:* **980-986,** 983 (ill.)
Geneva Protocol   *2:* 289
Genital herpes   *9:* 1735
Genital warts   *9:* 1735, 1737
Geocentric theory   *3:* 574
Geologic map   *5:* **986-989,** 988 (ill.)
Geologic time   *5:* **990-993**
Geologic time scale   *5:* 988
Geology   *5:* **993-994,** 944 (ill.)
Geometry   *5:* **995-999**
Geothermal energy   *1:* 116
Gerbert of Aurillac   *7:* 1396
Geriatrics   *5:* 999
Germ warfare.   *See* **Biological warfare**
Germanium   *2:* 401
Gerontology   *5:* **999**
Gestalt psychology   *8:* 1595
Gibberellin   *6:* 1051
Gilbert, William   *6:* 1212
Gillies, Harold Delf   *8:* 1529
Gills   *5:* 877
Glacier   *5:* **1000-1003,** 1002 (ill.)
Glaisher, James   *2:* 263
Glass   *5:* **1004-1006,** 1004 (ill.)
Glenn, John   *9:* 1779
Gliders   *1:* 77
Global Biodiversity Strategy   *2:* 283
Global climate   *5:* **1006-1009**
Globular clusters   *9:* 1802, 1808
Glucose   *2:* 388
Gluons   *10:* 1831
Glutamate   *9:* 1720
Glycerol   *1:* 91
Glycogen   *2:* 389
Gobi Desert   *1:* 199
Goddard, Robert H.   *9:* 1695 (ill.)

Goiter   *6:* 1220
Gold   *8:* 1566-1569
Goldberger, Joseph   *6:* 1219
Golden-brown algae   *1:* 93
Golgi body   *3:* 433
Gondwanaland   *1:* 149
Gonorrhea   *9:* 1735, 1736
Gorillas   *8:* 1572
Gould, Stephen Jay   *1:* 32
Graphs and graphing   *5:* **1009-1011**
Grasslands   *2:* 296
Gravitons   *10:* 1831
Gravity and gravitation   *5:* **1012-1016,** 1014 (ill.)
Gray, Elisha   *10:* 1867
Great Barrier Reef   *2:* 240
Great Dividing Range   *2:* 240
Great Lakes   *6:* 1159
Great Plains   *4:* 682, *7:* 1356
Great Red Spot (Jupiter)   *6:* 1149, 1150 (ill.)
Great Rift Valley   *1:* 49, 51
Great White Spot (Saturn)   *9:* 1709
Green algae   *1:* 94
Green flashes   *2:* 224
Greenhouse effect   *2:* 285, 393, *5:* 1003, **1016-1022,** 1020 (ill.), *8:* 1554, *10:* 1965
Gregorian calendar   *2:* 373, 375
Grissom, Virgil   *9:* 1779
Growth hormone   *5:* 797
Growth rings (trees)   *4:* 619
Guiana Highlands   *9:* 1772
Guided imagery   *1:* 119
Gum disease   *4:* 630
Guth, Alan   *2:* 276
Gymnophions   *1:* 137
Gymnosperms   *9:* 1729
Gynecology   *5:* **1022-1024,** 1022 (ill.)
Gyroscope   *5:* **1024-1025,** 1024 (ill.)

H

H.M.S. *Challenger* 7: *1413*
Haber process   *7:* 1346
Haber, Fritz   *7:* 1346
Hadley, George   *2:* 218
Hahn, Otto   *7:* 1361
Hale, Alan   *3:* 529
Hale-Bopp comet   *3:* 529

Index

Hales, Stephen *2:* 337
Half-life *6:* **1027**
Halite *6:* 1096
Hall, Charles M. *1:* 124, *4:* 757
Hall, Chester Moore *10:* 1871
Hall, John *7:* 1237
Halley, Edmond *7:* 1262, *10:* 2020
Halley's comet *3:* 528
Hallucinogens *6:* **1027-1030**
Haloes *2:* 224
Halogens *6:* **1030-1036**
Hand tools *6:* **1036-1037,** 1036 (ill.)
Hard water *9:* 1757
Hargreaves, James *6:* 1098
Harmonices mundi *3:* 425
Harmonics *5:* 925
Hart, William Aaron *7:* 1320
Harvestmen (spider) *1:* 171, 170 (ill.)
Harvey, William *1:* 139, *2:* 292
Hazardous waste *10:* 2006-2007, 2006 (ill.)
HDTV *10:* 1879
Heart *6:* **1037-1043,** 1041 (ill.), 1042 (ill.)
Heart attack *6:* 1043
Heart diseases *3:* 470, *6:* 1040
Heart transplants *10:* 1926
Heart, measure of electrical activity. *See* **Electrocardiogram**
Heartburn *4:* 656
Heat *6:* **1043-1046**
Heat transfer *6:* 1044
Heat, measurement of. *See* **Calorie**
Heisenberg, Werner *8:* 1609
Heliocentric theory *3:* 574
Helium *7:* 1349
Helminths *8:* 1471 (ill.)
Hemiptera *6:* 1105
Hemodialysis *5:* 841
Henbury Craters *2:* 240
Henry, Joseph *4:* 761, *10:* 1865
Herbal medicine *1:* 120
Herbicides *1:* 54-59
Herculaneum *5:* 828
Heredity *7:* 1246
Hermaphroditism *9:* 1667
Heroin *1:* 32, 34
Herophilus *1:* 138
Héroult, Paul *1:* 124
Herpes *9:* 1737
Herschel, John *2:* 277

Herschel, William *2:* 277, *10:* 1871, 1952
Hertz, Heinrich *6:* 1188, *8:* 1502, 1626
Hess, Henri *3:* 525
Hevelius, Johannes *7:* 1296
Hewish, Antony *7:* 1340
Hibernation *6:* **1046-1048,** 1047 (ill.)
Himalayan Mountains *1:* 194, 197, *7:* 1301
Hindbrain *2:* 340
Hindenburg *1:* 76
Hindu-Arabic number system *1:* 178, *7:* 1396, *10:* 2047
Hippocrates *2:* 348
Histamine *6:* 1085
Histology *1:* 141
Historical concepts *6:* 1186
HIV (human immunodeficiency virus) *1:* 70, 72 (ill.), *8:* 1583
Hodgkin's lymphoma *6:* 1201
Hoffmann, Felix *1:* 6
Hofstadter, Robert *7:* 1339
Hogg, Helen Sawyer *10:* 1964
Holistic medicine *1:* 120
Holland, John *10:* 1835
Hollerith, Herman *3:* 549
Holmium *6:* 1163
Holograms and holography *6:* **1048-1050,** 1049 (ill.)
Homeopathy *1:* 120
Homeostasis *8:* 1516, 1517
Homo erectus *6:* 1058
Homo ergaster *6:* 1058
Homo habilis *6:* 1058
Homo sapiens *6:* 1055, 1058-1059
Homo sapiens sapiens *6:* 1059
Hooke, Robert *1:* 140, *4:* 731
Hooke's law *4:* 731
Hopewell mounds *7:* 1301
Hopkins, Frederick G. *10:* 1982
Hopper, Grace *3:* 551
Hormones *6:* **1050-1053**
Horticulture *6:* **1053-1054,** 1053 (ill.)
HTTP *6:* 1128
Hubble Space Telescope *9:* 1808, *10:* 1873, 1872 (ill.)
Hubble, Edwin *2:* 275, *7:* 1328, *9:* 1655, 1810
Human evolution *6:* **1054-1060,** 1057 (ill.), 1059 (ill.)
Human Genome Project *6:* **1060-**

Index

1068, 1062 (ill.), 1065 (ill.), 1066 (ill.)
Human-dominated biomes   *2:* 302
Humanistic psychology   *8:* 1596
Humason, Milton   *9:* 1655
Hurricanes   *3:* 610
Hutton, James   *10:* 1947
Huygens, Christiaan   *6:* 1187, *9:* 1711
Hybridization   *2:* 310
Hydrocarbons   *7:* 1430-1431
Hydrogen   *6:* **1068-1071,** 1068 (ill.)
Hydrologic cycle   *6:* **1071-1075,** 1072 (ill.), 1073 (ill.)
Hydropower   *1:* 113
Hydrosphere   *2:* 305
Hydrothermal vents   *7:* 1418, 1417 (ill.)
Hygrometer   *10:* 2020
Hypertension   *3:* 484
Hypertext   *6:* 1128
Hypnotherapy   *1:* 119
Hypotenuse   *10:* 1932
Hypothalamus   *2:* 342, 343
Hypothesis   *9:* 1723

I

Icarus   *1:* 74
Ice ages   *6:* **1075-1078,** 1077 (ill.)
Icebergs   *6:* **1078-1081,** 1080 (ill.), 1081 (ill.)
Idiot savants.   *See* **Savants**
IgE   *1:* 109
Igneous rock   *9:* 1702
Ileum   *4:* 656
Imaginary numbers   *6:* **1081-1082**
Immune system   *1:* 108, *6:* **1082-1087**
Immunization   *1:* 161, *10:* 1060-1960
Immunoglobulins   *1:* 159
Imprinting   *2:* 272
In vitro fertilization   *4:* 791
Incandescent light   *6:* **1087-1090,** 1089 (ill.)
Inclined plane   *6:* 1207
Indian peninsula   *1:* 197
Indicator species   *6:* **1090-1092,** 1091 (ill.)
Indium   *1:* 126
Induction   *4:* 760

Industrial minerals   *6:* **1092-1097**
Industrial Revolution   *1:* 28, *3:* 523, *6:* 1193, **1097-1100,** *7:* 1236, *9:* 1817
   automation   *2:* 242
   effect on agriculture   *1:* 63
   food preservation   *5:* 892
Infantile paralysis.   *See* **Poliomyelitis**
Infants, sudden death.   *See* **Sudden infant death syndrome (SIDS)**
Inflationary theory   *2:* 275, 276
Influenza   *4:* 672, *6:* 1084, *10:* 1978, 1979-1981
Infrared Astronomical Satellite   *9:* 1808
Infrared astronomy   *6:* **1100-1103,** 1102 (ill.)
Infrared telescopes   *6:* 1101
Ingestion   *4:* 653
Inheritance, laws of.   *See* **Mendelian laws of inheritance**
Insecticides   *1:* 67
Insects   *6:* **1103-1106,** 1104 (ill.)
Insomnia   *9:* 1747
Insulin   *3:* 474, *4:* 638
Integers   *1:* 180
Integral calculus   *2:* 372
Integrated circuits   *6:* **1106-1109,** 1108 (ill.), 1109 (ill.)
Integumentary system   *6:* **1109-1112,** 1111 (ill.)
Interference   *6:* **1112-1114,** 1113 (ill.)
Interferometer   *6:* 1115 (ill.), 1116
Interferometry   *10:* 1874, *6:* **1114-1116,** 1115 (ill.), 1116 (ill.)
Interferon   *6:* 1084
Internal-combustion engines   *6:* **1117-1119,** 1118 (ill.)
International Space Station   *9:* 1788
International System of Units   *2:* 376
International Ultraviolet Explorer   *10:* 1946, *6:* **1120-1123,** 1122 (ill.)
Internet   *6:* **1123-1130,** 1127 (ill.)
Interstellar matter   *6:* **1130-1133,** 1132 (ill.)
Invertebrates   *6:* **1133-1134,** 1134 (ill.)
Invertebrates, age of   *8:* 1461
Io   *6:* 1148, 1149
Iodine   *6:* 1035
Ionic bonding   *3:* 455
Ionization   *6:* **1135-1137**
Ionization energy   *6:* 1135

Index

Ions *4:* 733
Iron *10:* 1915-1918
Iron lung *8:* 1548 (ill.)
Iron manufacture *6:* 1098
Irrational numbers *1:* 180, 181
Isaacs, Alick *6:* 1084
Islands *3:* 500, *6:* **1137-1141,** 1139 (ill.)
Isotopes *6:* **1141-1142,** *7:* 1241
IUE. *See* **International Ultraviolet Explorer**

J

Jackals *2:* 385
Jacquet-Droz, Henri *9:* 1691
Jacquet-Droz, Pierre *9:* 1691
James, William *8:* 1594
Jansky, Karl *8:* 1635
Java man *6:* 1058
Jefferson, Thomas *7:* 1300
Jejunum *4:* 655
Jenner, Edward *1:* 161, *10:* 1957
Jet engines *6:* **1143-1146,** 1143 (ill.), 1145 (ill.)
Jet streams *2:* 221, *4:* 783, *7:* 1293
Jones, John *8:* 1529
Joule *6:* 1045
Joule, James *10:* 1885
Joule-Thomson effect *3:* 597
Jupiter (planet) *6:* **1146-1151,** 1147 (ill.), 1150 (ill.)

K

Kangaroos and wallabies *6:* **1153-1157,** 1155 (ill.)
Kant, Immanuel *9:* 1765
Kay, John *6:* 1097
Kelvin scale *10:* 1882
Kelvin, Lord. *See* **Thomson, William**
Kenyanthropus platyops *6:* 1056
Kepler, Johannes *3:* 425, 574, *7:* 1426
Keratin *6:* 1110
Kettlewell, Henry Bernard David *1:* 28
Kidney dialysis *4:* 645
Kidney stones *5:* 841
Kilimanjaro, Mount *5:* 1000
Kinetic theory of matter *7:* 1243

King, Charles G. *6:* 1219
Klein bottle *10:* 1899
Knowing. *See* **Cognition**
Klein, Felix *10:* 1899
Koch, Robert *2:* 292
Köhler, Wolfgang *8:* 1595
Kraepelin, Emil *9:* 1718
Krakatoa *10:* 1998
Krypton *7:* 1349, 1352
Kuiper Disk *3:* 530
Kuiper, Gerald *3:* 530, *7:* 1333
Kwashiorkor *6:* 1218, *7:* 1403

L

La Niña *4:* 782
Lacrimal gland *5:* 852
Lactose *2:* 388
Lager *2:* 354
Lake Baikal *1:* 198
Lake Huron *6:* 1162 (ill.)
Lake Ladoga *5:* 824
Lake Michigan *7:* 1354 (ill.)
Lake Superior *6:* 1159
Lake Titicaca *6:* 1159
Lakes *6:* **1159-1163,** 1161 (ill.), 1162 (ill.)
Lamarck, Jean-Baptiste *1:* 28
Lambert Glacier *1:* 149
Laminar flow *1:* 40
Landfills *10:* 2007, 2008 (ill.)
Language *3:* 515
Lanthanides *6:* **1163-1166**
Lanthanum *6:* 1163
Laplace, Pierre-Simon *2:* 323, *9:* 1765
Large intestine *4:* 656
Laryngitis *9:* 1681
Laser eye surgery *8:* 1617
Lasers *6:* **1166-1168,** 1168 (ill.)
LASIK surgery *8:* 1617
Laurentian Plateau *7:* 1355
Lava *10:* 1995
Lavoisier, Antoine Laurent *3:* 465, 524, *6:* 1069, *7:* 1444
Law of conservation of energy *3:* 555, *10:* 1885
Law of conservation of mass/matter *3:* 554, *5:* 816
Law of conservation of momentum *7:* 1290

Index

Law of dominance *7:* 1249
Law of electrical force *3:* 579
Law of independent assortment *7:* 1249
Law of planetary motion *3:* 425
Law of segregation *7:* 1249
Law of universal gravitation *3:* 426, *7:* 1427
Lawrence, Ernest Orlando *8:* 1479
Laws of motion *3:* 426, *6:* **1169-1171,** *7:* 1235, 1426
Le Verrier, Urbain *7:* 1330
Lead *2:* 402-403
Leakey, Louis S. B. *6:* 1058
Learning disorders *4:* 690
Leaf *6:* **1172-1176,** 1172 (ill.), 1174 (ill.)
Leavitt, Henrietta Swan *10:* 1964
Leclanché, Georges *2:* 269
LED (light-emitting diode) *6:* **1176-1179,** 1177 (ill.), 1178 (ill.)
Leeuwenhoek, Anton van *2:* 253, *6:* 1184, *8:* 1469
Legionella pneumophilia *6:* 1181, 1182
Legionnaire's disease *6:* **1179-1184,** 1182 (ill.)
Leibniz, Gottfried Wilhelm *2:* 371, 372, *7:* 1242
Lemaître, Georges-Henri *3:* 576
Lemurs *8:* 1572
Lenoir, Jean-Joseph Éttien *6:* 1119
Lenses *6:* **1184-1185,** 1184 (ill.)
Lentic biome *2:* 298
Leonid meteors *7:* 1263
Leonov, Alexei *9:* 1779
Leopards *5:* 860
Leptons *10:* 1830
Leucippus *2:* 226
Leukemia *2:* 380
Leukocytes *2:* 328
Lever *6:* 1205, 1206 (ill.)
Levy, David *6:* 1151
Lewis, Gilbert Newton *1:* 15
Liber de Ludo Aleae *8:* 1576
Lice *8:* 1474
Life, origin of *4:* 702
Light *6:* 1087-1090, **1185-1190**
Light, speed of *6:* 1190
Light-year *6:* **1190-1191**
Lightning *10:* 1889, 1889 (ill.)
Limbic system *2:* 345
Liming agents *1:* 66

Lind, James *6:* 1218, *10:* 1981
Lindenmann, Jean *6:* 1084
Linear acceleration *1:* 4
Linear accelerators *8:* 1477
Linnaeus, Carolus *2:* 292, 337
Lions *5:* 860
Lipids *6:* **1191-1192,** *7:* 1400
Lippershey, Hans *10:* 1869
Liquid crystals *7:* 1244-1245
Liquor. See **Alcohol (liquor)**
Lister, Joseph *1:* 165
Lithium *1:* 100
Lithosphere *8:* 1535, 1536
Litmus test *8:* 1496
Locks (water) *6:* **1192-1195,** 1193 (ill.)
Logarithms *6:* **1195**
Long, Crawford W. *1:* 143
Longisquama insignis *2:* 312
Longitudinal wave *10:* 2015
Lord Kelvin. See **Thomson, William**
Lorises *8:* 1572
Lotic *2:* 299
Lotic biome *2:* 299
Lowell, Percival *8:* 1539
Lowry, Thomas *1:* 15
LSD *6:* 1029
Lucy (fossil) *6:* 1056, 1057 (ill.)
Luminescence *6:* **1196-1198,** 1196 (ill.)
Luna *7:* 1296
Lunar eclipses *4:* 725
Lunar Prospector *7:* 1297
Lung cancer *9:* 1682
Lungs *9:* 1679
Lunisolar calendar *2:* 374
Lutetium *6:* 1163
Lymph *6:* 1199
Lymph nodes *6:* 1200
Lymphatic system *6:* **1198-1202**
Lymphocytes *2:* 329, *6:* 1085, 1200 (ill.)
Lymphoma *2:* 380, *6:* 1201
Lysergic acid diethylamide. See **LSD**

M

Mach number *5:* 883
Mach, L. *6:* 1116
Machines, simple *6:* **1203-1209,** 1206 (ill.), 1208 (ill.)

Index

Mackenzie, K. R.   *6:* 1035
Magellan   *10:* 1966
Magma   *10:* 1995
Magnesium   *1:* 103
Magnetic fields   *4:* 759
Magnetic fields, stellar.   *See* **Stellar magnetic fields**
Magnetic recording/audiocassette   *6:* **1209-1212**, 1209 (ill.), 1211 (ill.)
Magnetic resonance imaging   *2:* 304
Magnetism   *6:* **1212-1215**, 1214 (ill.)
Malnutrition   *6:* **1216-1222**, 1221 (ill.)
Malpighi, Marcello   *1:* 139
Mammals   *6:* **1222-1224**, 1223 (ill.)
Mammals, age of   *8:* 1462
Mangrove forests   *5:* 909
Manhattan Project   *7:* 1365, 1380
Manic-depressive illness   *4:* 631
Mantle   *4:* 710
Manufacturing.   *See* **Mass production**
MAP (Microwave Anisotroy Probe)   *2:* 276
Maps and mapmaking.   *See* **Cartography**
Marasmus   *6:* 1218
Marconi, Guglielmo   *8:* 1626
Marie-Davy, Edme Hippolyte   *10:* 2021
Marijuana   *6:* **1224-1227**, 1226 (ill.), 1227 (ill.)
Marine biomes   *2:* 299
Mariner 10   *7:* 1250
Mars (planet)   *6:* **1228-1234**, 1228 (ill.), 1231 (ill.), 1232 (ill.)
Mars Global Surveyor   *6:* 1230
Mars Pathfinder   *6:* 1232
Marshes   *10:* 2025
Maslow, Abraham   *8:* 1596
Mass   *7:* **1235-1236**
Mass production   *7:* **1236-1239**, 1238 (ill.)
Mass spectrometry   *7:* **1239-1241**, 1240 (ill.), *8:* 1604
Mastigophora   *8:* 1592
Mathematics   *7:* **1241-1242**
   imaginary numbers   *6:* 1081
   logarithms   *6:* 1195
   multiplication   *7:* 1307
   number theory   *7:* 1393
   probability theory   *8:* 1575
   proofs   *8:* 1578
   statistics   *9:* 1810
   symbolic logic   *10:* 1859
   topology   *10:* 1897-1899
   trigonometry   *10:* 1931-1933
   zero   *10:* 2047
Matter, states of   *7:* **1243-1246**, 1243 (ill.)
Maxwell, James Clerk   *4:* 760, 767, *6:* 1213, *8:* 1626
Maxwell's equations   *4:* 760
McKay, Frederick   *5:* 889
McKinley, Mount   *7:* 1302
McMillan, Edwin   *1:* 24
Measurement.   *See* **Units and standards**
Mechanoreceptors   *8:* 1484
Meditation   *1:* 119
Medulla oblongata   *2:* 340
Meiosis   *9:* 1666
Meissner effect   *10:* 1851 (ill.)
Meitner, Lise   *7:* 1362
Mekong River   *1:* 200
Melanin   *6:* 1110
Melanomas   *2:* 380
Memory   *2:* 344, *3:* 515
Mendel, Gregor   *2:* 337, *4:* 786, *7:* 1247
Mendeleev, Dmitry   *4:* 777, *8:* 1487
Mendelian laws of inheritance   *7:* **1246-1250**, 1248 (ill.)
Meninges   *2:* 342
Menopause   *1:* 59, *2:* 410, *5:* 800, 1020
Menstruation   *1:* 59, *5:* 800, 1020, *8:* 1599
Mental illness, study and treatment of.   *See* **Psychiatry**
Mercalli scale   *4:* 704
Mercalli, Guiseppe   *4:* 704
Mercury   *10:* 1921-1923, 1922 (ill.)
Mercury (planet)   *7:* **1250-1255**, 1251 (ill.), 1252 (ill.)
Mercury barometers   *2:* 265
Méré, Chevalier de   *8:* 1576
Mescaline   *6:* 1029
Mesosphere   *2:* 213
Mesozoic era   *5:* 990, *8:* 1462
Metabolic disorders   *7:* **1254-1255**, 1254 (ill.), 1257
Metabolism   *7:* **1255-1259**
Metalloids   *7:* 1348
Metamorphic rocks   *9:* 1705
Metamorphosis   *7:* **1259-1261**

Meteorograph *2:* 215
Meteors and meteorites *7:* **1262-1264**
Methanol *1:* 89
Metric system *7:* **1265-1268,** *10:* 1949
Mettauer, John Peter *8:* 1529
Meyer, Julius Lothar *4:* 777, *8:* 1487
Michell, John *2:* 323
Michelson, Albert A. *6:* 1114, *6:* 1187
Microwave Anisotropy Probe (MAP) *2:* 276
Microwave communication *7:* **1268-1271,** 1270 (ill.)
Microwaves *4:* 765
Mid-Atlantic Ridge *7:* 1303, 1409
Midbrain *2:* 340
Middle East *1:* 196
Mifepristone *3:* 565
Migraine *2:* 349, 350
Migration (animals) *7:* **1271-1273,** 1272 (ill.)
Millennium *2:* 375
Millikan, Robert Andrew *4:* 771
Minerals *6:* 1092-1097, *7:* 1401, **1273-1278,** 1276 (ill.), 1277 (ill.)
Mining *7:* **1278-1282,** 1281 (ill.)
Mir *9:* 1781
Mirages *2:* 222
Miranda *10:* 1954
Misch metal *6:* 1165
Missiles. *See* **Rockets and missiles**
Mission *9:* 1787
Mississippi River *7:* 1355
Mississippian earthern mounds *7:* 1301
Missouri River *7:* 1355
Mitchell, Mount *7:* 1356
Mites *1:* 170
Mitochondira *3:* 436
Mitosis *1:* 133, *9:* 1665
Mobile telephones *3:* 441
Möbius strip *10:* 1899
Möbius, Augustus Ferdinand *10:* 1899
Model T (automobile) *7:* 1237, 1238
Modulation *8:* 1627
Moho *4:* 709
Mohorovičiá discontinuity *4:* 709
Mohorovičiá, Andrija *4:* 709
Mohs scale *1:* 3
Mohs, Friedrich *1:* 3
Mole (measurement) *7:* **1282-1283**
Molecular biology *7:* **1283-1285**
Molecules *7:* **1285-1288,** 1286 (ill.)

Mollusks *7:* **1288-1290,** 1289 (ill.)
Momentum *7:* **1290-1291**
Monkeys *8:* 1572
Monoclonal antibodies *1:* 162, *2:* 311
Monocytes *2:* 329
Monoplacophora *7:* 1289
Monosaccharides *2:* 388
Monotremes *6:* 1224
Monsoons *7:* **1291-1294**
Mont Blanc *5:* 827
Montgolfier, Jacques *2:* 262
Montgolfier, Joseph *2:* 262
Moon *7:* **1294-1298,** 1295 (ill.), 1297 (ill.)
  affect on tides *10:* 1890
  Apollo 11 *9: 1779*
Morley, Edward D. *6:* 1187
Morphine *1:* 32, 33
Morse code *10:* 1866
Morse, Samuel F. B. *10:* 1865
Morton, William *1:* 143
Mosquitoes *6:* 1106, *8:* 1473
Motion, planetary, laws of *7:* 1426
Motors, electric. *See* **Electric motors**
Mounds, earthen *7:* **1298-1301,** 1299 (ill.)
Mount Ararat *1:* 197
Mount Cameroon *1:* 51
Mount Elbert *7:* 1357
Mount Elbrus *5:* 823
Mount Everest *1:* 194
Mount Kilimanjaro *1:* 53 (ill.), *5:* 1003, *7:* 1303
Mount McKinley *7:* 1302, 1354
Mount Mitchell *7:* 1356
Mount Robson *7:* 1357
Mount St. Helens *10:* 1996, 1998 (ill.)
Mountains *7:* **1301-1305,** 1304 (ill.)
Movable bridges *2:* 358
mRNA *7:* 1285
Müller, Karl Alex *10:* 1851
Multiple personality disorder *7:* **1305-1307**
Multiplication *7:* **1307-1309**
Muscular dystrophy *7:* 1313, 1337
Muscular system *7:* **1309-1313,** 1311 (ill.), 1312 (ill.), 1313 (ill.)
Mushrooms *6:* 1028
Mutation *7:* **1314-1317,** 1316 (ill.)
Mysophobia *8:* 1497

Index

N

Napier, John *6:* 1195
Narcolepsy *9:* 1748
Narcotics *1:* 32
Natural gas *7:* **1319-1321**
Natural language processing *1:* 189
Natural numbers *1:* 180, *7:* **1321-1322**
Natural selection *1:* 29-30, *2:* 292, *5:* 834, 837-839
Natural theology *1:* 27
Naturopathy *1:* 122
Nautical archaeology *7:* **1323-1327,** 1325 (ill.), 1326 (ill.)
Nautilus *10:* 1835
Navigation (animals) *7:* 1273
Neanderthal man *6:* 1059
Neap tides *10:* 1892
NEAR Shoemaker *1:* 203-204, *9:* 1787
Nearsightedness *5:* 851
Nebula *7:* **1327-1330,** 1329 (ill.)
Nebular hypothesis *9:* 1765
Negative reinforcement. *See* **Reinforcement, positive and negative**
Nematodes *8:* 1471
Neodymium *6:* 1163
Neolithic Age *1:* 62
Neon *7:* 1349, 1352
Neptune (planet) *7:* **1330-1333,** 1331 (ill.)
Nereid *7:* 1333
Nerve nets *7:* 1333
Nervous system *7:* **1333-1337,** 1335 (ill.)
Neurons *7:* 1333 (ill.)
Neurotransmitters *1:* 128, *2:* 350, *4:* 631, *7:* 1311, *9:* 1720
Neutralization *1:* 15
Neutrinos *10:* 1832, 1833
Neutron *2:* 235, *7:* **1337-1339,** *10:* 1830, 1832
Neutron stars *7:* **1339-1341,** 1341 (ill.)
Neutrophils *2:* 329
Newcomen, Thomas *9:* 1818
Newton, Isaac *1:* 4, *5:* 1012, *6:* 1184, *7:* 1242
  calculus *2:* 372
  corpsucular theory of light *6:* 1187
  laws of motion *6:* 1169-1171, *7:* 1427

reflector telescope *10:* 1871
Ngorongoro Crater *1:* 52
Niacin. *See* **Vitamin B3**
Nicotine *1:* 34, *3:* 478
Night blindness *6:* 1220
Nile River *9:* 1686
Nipkow, Paul *10:* 1875
Nitrification *7:* 1343, 1344
Nitrogen *7:* 1344, 1345 (ill.)
Nitrogen cycle *7:* **1342-1344,** 1342 (ill.)
Nitrogen family *7:* **1344-1349,** 1345 (ill.)
Nitrogen fixation *7:* 1342
Nitroglycerin *5:* 844
Nitrous oxide *1:* 142
Nobel, Alfred *5:* 845
Noble gases *7:* **1349-1352,** 1350 (ill.)
Non-Hodgkin's lymphoma *6:* 1201
Nondestructive testing *10:* 2037
North America *7:* **1352-1358,** 1353 (ill.), 1354 (ill.), 1357 (ill.)
Novas *7:* **1359-1360,** 1359 (ill.)
NSFNET *6:* 1127
Nuclear fission *7:* **1361-1366,** 1365 (ill.), 1381
Nuclear fusion *7:* **1366-1371,** 1370 (ill.)
Nuclear medicine *7:* **1372-1374**
Nuclear Non-Proliferation Treaty *7:* 1387
Nuclear power *1:* 113, *7:* **1374-1381,** 1376 (ill.), 1379 (ill.), *10:* 1836
Nuclear power plant *7:* 1374, 1379 (ill.)
Nuclear reactor *7:* 1365
Nuclear waste management *7:* 1379
Nuclear weapons *7:* **1381-1387**
Nucleic acids *7:* **1387-1392,** 1391 (ill.), 1392 (ill.)
Nucleotides *3:* 473, *7:* 1388
Nucleus (cell) *3:* 434
Number theory *7:* 1322, **1393-1395**
Numbers, imaginary *6:* 1081
Numbers, natural *7:* 1321
Numeration systems *7:* **1395-1399**
Nutrient cycle *2:* 307
Nutrients *7:* 1399
Nutrition *6:* 1216, *7:* **1399-1403,** 1402 (ill.)
Nylon *1:* 186, *8:* 1533

O

Oberon *10:* 1954
Obesity *4:* 716
Obsession *7:* **1405-1407**
Obsessive-compulsive disorder *7:* 1405
Obsessive-compulsive personality disorder *7:* 1406
Occluded fronts *1:* 82
Ocean *7:* **1407-1411,** 1407 (ill.)
Ocean currents *3:* 604-605
Ocean ridges *7:* 1410 (ill.)
Ocean zones *7:* **1414-1418**
Oceanic archaeology. *See* **Nautical archaeology**
Oceanic ridges *7:* 1409
Oceanography *7:* **1411-1414,** 1412 (ill.), 1413 (ill.)
Octopus *7:* 1289
Oersted, Hans Christian *1:* 124, *4:* 760, 766, *6:* 1212
Offshore drilling *7:* 1421
Ohio River *7:* 1355
Ohm (O) *4:* 738
Ohm, Georg Simon *4:* 738
Ohm's law *4:* 740
Oil drilling *7:* **1418-1422,** 1420 (ill.)
Oil pollution *7:* 1424
Oil spills *7:* **1422-1426,** 1422 (ill.), 1425 (ill.)
Oils *6:* 1191
Olduvai Gorge *6:* 1058
Olfaction. *See* **Smell**
On the Origin of Species by Means of Natural Selection *6:* 1054
On the Structure of the Human Body *1:* 139
O'Neill, J. A. *6:* 1211
Onnes, Heike Kamerlingh *10:* 1850
Oort cloud *3:* 530
Oort, Jan *8:* 1637
Open clusters *9:* 1808
Open ocean biome *2:* 299
Operant conditioning *9:* 1658
Ophediophobia *8:* 1497
Opiates *1:* 32
Opium *1:* 32, 33
Orangutans *8:* 1572, 1574 (ill.)
Orbit *7:* **1426-1428**
Organ of Corti *4:* 695
Organic chemistry *7:* **1428-1431**
Organic families *7:* 1430
Organic farming *7:* **1431-1434,** 1433 (ill.)
Origin of life *4:* 702
Origins of algebra *1:* 97
Orizaba, Pico de *7:* 1359
Orthopedics *7:* **1434-1436**
Oscilloscopes *10:* 1962
Osmosis *4:* 652, *7:* **1436-1439,** 1437 (ill.)
Osmotic pressure *7:* 1436
Osteoarthritis *1:* 181
Osteoporosis *9:* 1742
Otitis media *4:* 697
Otosclerosis *4:* 697
Ovaries *5:* 800
Oxbow lakes *6:* 1160
Oxidation-reduction reactions *7:* **1439-1442,** *9:* 1648
Oxone layer *7:* 1452 (ill.)
Oxygen family *7:* **1442-1450,** 1448 (ill.)
Ozone *7:* **1450-1455,** 1452 (ill.)
Ozone depletion *1:* 48, *8:* 1555
Ozone layer *7:* 1451

P

Packet switching *6:* 1124
Pain *7:* 1336
Paleoecology *8:* **1457-1459,** 1458 (ill.)
Paleontology *8:* **1459-1462,** 1461 (ill.)
Paleozoic era *5:* 990, *8:* 1461
Paleozoology *8:* 1459
Panama Canal *6:* 1194
Pancreas *4:* 655, *5:* 798
Pangaea *8:* 1534, 1536 (ill.)
Pap test *5:* 1020
Papanicolaou, George *5:* 1020
Paper *8:* **1462-1467,** 1464 (ill.), 1465 (ill.), 1466 (ill.)
Papyrus *8:* 1463
Paracelsus, Philippus Aureolus *1:* 84
Parasites *8:* **1467-1475,** 1471 (ill.), 1472 (ill.), 1474 (ill.)
Parasitology *8:* 1469
Parathyroid glands *5:* 798
Paré, Ambroise *8:* 1580, *10:* 1855
Parkinson's disease *1:* 62
Parsons, Charles A. *9:* 1820
Particle accelerators *8:* **1475-1482,**

Index

1478 (ill.), 1480 (ill.)
Particulate radiation *8:* 1620
Parturition. *See* **Birth**
Pascal, Blaise *2:* 370, *8:* 1576
Pascaline *2:* 371
Pasteur, Louis *1:* 161, 165, *2:* 292, *10:* 1958, 1990, 1959 (ill.)
Pavlov, Ivan P. *9:* 1657, *10:* 1990
Pelagic zone *7:* 1415
Pellagra *6:* 1219
Penicillin *1:* 155, 156, 157 (ill.)
Peppered Moth *1:* 28
Perception *8:* **1482-1485,** 1483 (ill.)
Periodic function *8:* **1485-1486,** 1485 (ill.)
Periodic table *4:* 777, 778 (ill.), *8:* 1489, **1486-1490,** 1489 (ill.), 1490 (ill.)
Periscopes *10:* 1835
Perrier, C. *4:* 775, *10:* 1913
Perseid meteors *7:* 1263
Persian Gulf War *3:* 462, *7:* 1425
Pesticides *1:* 67, 67 (ill.), 68 (ill.), *4:* 619-622
PET scans *2:* 304, *8:* 1640
Petroglyphs and pictographs *8:* **1491-1492,** 1491 (ill.)
Petroleum *7:* 1418, 1423, *8:* **1492-1495**
Peyote *6:* 1029
Pfleumer, Fritz *6:* 1211
pH *8:* **1495-1497**
Phaeophyta *1:* 95
Phages *10:* 1974
Phenothiazine *10:* 1906
Phenylketonuria *7:* 1254
Phloem *6:* 1175, *8:* 1523
Phobias *8:* **1497-1498**
Phobos *6:* 1229
Phosphates *6:* 1095
Phosphorescence *6:* 1197
Phosphorus *7:* 1347
Photochemistry *8:* **1498-1499**
Photocopying *8:* **1499-1502,** 1500 (ill.), 1501 (ill.)
Photoelectric cell *8:* 1504
Photoelectric effect *6:* 1188, *8:* **1502-1505**
Photoelectric theory *8:* 1503
Photoreceptors *8:* 1484
Photosphere *10:* 1846
Photosynthesis *2:* 306, 388, 391, *8:* **1505-1507**
Phototropism *6:* 1051, *8:* **1508-1510,** 1508 (ill.)
Physical therapy *8:* **1511-1513,** 1511 (ill.)
Physics *8:* **1513-1516**
Physiology *8:* **1516-1518**
Phytoplankton *8:* 1520, 1521
Pia mater *2:* 342
Piazzi, Giuseppe *1:* 201
Pico de Orizaba *7:* 1359
Pictographs, petroglyphs and *8:* 1491-1492
Pigments, dyes and *4:* 686-690
Pineal gland *5:* 798
PKU (phenylketonuria) *7:* 1254
Place value *7:* 1397
Plages *10:* 1846
Plague *8:* **1518-1520,** 1519 (ill.)
Planck's constant *8:* 1504
Plane *6:* 1036 (ill.), 1207
Planetary motion, laws of *7:* 1426
Plankton *8:* **1520-1522**
Plant behavior *2:* 270
Plant hormones *6:* 1051
Plants *1:* 91, *2:* 337, 388, 392, *8:* 1505, **1522-1527,** 1524 (ill.), 1526 (ill.)
Plasma *2:* 326, *7:* 1246
Plasma membrane *3:* 432
Plastic surgery *8:* **1527-1531,** 1530 (ill.), *10:* 1857
Plastics *8:* **1532-1534,** 1532 (ill.)
Plastids *3:* 436
Plate tectonics *8:* **1534-1539,** 1536 (ill.), 1538 (ill.)
Platelets *2:* 329
Platinum *8:* 1566, 1569-1570
Pluto (planet) *8:* **1539-1542,** 1541 (ill.)
Pneumonia *9:* 1681
Pneumonic plague *8:* 1520
Poisons and toxins *8:* **1542-1546,** 1545 (ill.)
Polar and nonpolar bonds *3:* 456
Poliomyelitis *8:* **1546-1549,** 1548 (ill.), *10:* 1958
Pollination *5:* 880-882
Pollution *1:* 9, *8:* **1549-1558,** 1554 (ill.), 1557 (ill.)

Index

Pollution control *8:* **1558-1562,** 1559 (ill.), 1560 (ill.)
Polonium *7:* 1449, 1450
Polygons *8:* **1562-1563,** 1562 (ill.)
Polymers *8:* **1563-1566,** 1565 (ill.)
Polysaccharides *2:* 388
Pompeii *5:* 828, *10:* 1997
Pons, Stanley *7:* 1371
Pontiac Fever *6:* 1181
Pope Gregory XIII *2:* 373
Porpoises *3:* 448
Positive reinforcement. *See* **Reinforcement, positive and negative**
Positron *1:* 163, *4:* 772
Positron-emission tomography. *See* **PET scans**
Positrons *10:* 1832, 1834
Post-it™ notes *1:* 38
Post-traumatic stress disorder *9:* 1826
Potassium *1:* 102
Potassium salts *6:* 1095
Potential difference *4:* 738, 744
Pottery *3:* 447
Praseodymium *6:* 1163
Precambrian era *5:* 988, *8:* 1459
Precious metals *8:* **1566-1570,** 1568 (ill.)
Pregnancy, effect of alcohol on *1:* 87
Pregnancy, Rh factor in *9:* 1684
Pressure *8:* **1570-1571**
Priestley, Joseph *2:* 394, 404, *3:* 525, *7:* 1345, 1444
Primary succession *10:* 1837, 2026
Primates *8:* **1571-1575,** 1573 (ill.), 1574 (ill.)
Probability theory *8:* **1575-1578**
Procaryotae *2:* 253
Progesterone *5:* 800
Projectile motion. *See* **Ballistics**
Prokaryotes *3:* 429
Promethium *6:* 1163
Proof (mathematics) *8:* **1578-1579**
Propanol *1:* 91
Prosthetics *8:* **1579-1583,** 1581 (ill.), 1582 (ill.)
Protease inhibitors *8:* **1583-1586,** 1585 (ill.)
Proteins *7:* 1399, *8:* **1586-1589,** 1588 (ill.)
Protons *10:* 1830, 1832
Protozoa *8:* 1470, **1590-1592,** 1590 (ill.)

Psilocybin *6:* 1029
Psychiatry *8:* **1592-1594**
Psychoanalysis *8:* 1594
Psychology *8:* **1594-1596**
Psychosis *8:* **1596-1598**
Ptolemaic system *3:* 574
Puberty *8:* **1599-1601,** *9:* 1670
Pulley *6:* 1207
Pulsars *7:* 1340
Pyrenees *5:* 826
Pyroclastic flow *10:* 1996
Pyrrophyta *1:* 94
Pythagoras of Samos *8:* 1601
Pythagorean theorem *8:* **1601**
Pytheas *10:* 1890

Q

Qualitative analysis *8:* **1603-1604**
Quantitative analysis *8:* **1604-1607**
Quantum mechanics *8:* **1607-1609**
Quantum number *4:* 772
Quarks *10:* 1830
Quartz *2:* 400
Quasars *8:* **1609-1613,** 1611 (ill.)

R

Rabies *10:* 1958
Radar *8:* **1613-1615,** 1614 (ill.)
Radial keratotomy *8:* **1615-1618,** 1618 (ill.)
Radiation *6:* 1044, *8:* **1619-1621**
Radiation exposure *8:* **1621-1625,** 1623 (ill.), 1625 (ill.)
Radio *8:* **1626-1628,** 1628 (ill.)
Radio astronomy *8:* **1633-1637,** 1635 (ill.)
Radio waves *4:* 765
Radioactive decay dating *4:* 616
Radioactive fallout *7:* 1385, 1386
Radioactive isotopes *6:* 1142, *7:* 1373
Radioactive tracers *8:* **1629-1630**
Radioactivity *8:* **1630-1633**
Radiocarbon dating *1:* 176
Radiology *8:* **1637-1641,** 1640 (ill.)
Radionuclides *7:* 1372
Radiosonde *2:* 216
Radium *1:* 105

Index

Radon *7:* 1349, 1350
Rain forests *2:* 295, *8:* **1641-1645,** 1643 (ill.), 1644 (ill.)
Rainbows *2:* 222
Rainforests *8:* 1643 (ill.)
Ramjets *6:* 1144
Rare earth elements *6:* 1164
Rat-kangaroos *6:* 1157
Rational numbers *1:* 180
Rawinsonde *2:* 216
Reaction, chemical
Reaction, chemical *9:* **1647-1649,** 1649 (ill.)
Reality engine *10:* 1969, 1970
Reber, Grote *8:* 1635
Receptor cells *8:* 1484
Recommended Dietary Allowances *10:* 1984
Reconstructive surgery. *See* **Plastic surgery**
Recycling *9:* **1650-1653,** 1650 (ill.), 1651 (ill.), *10:* 2009
Red algae *1:* 94
Red blood cells *2:* 327, 328 (ill.)
Red giants *9:* **1653-1654**
Red tides *1:* 96
Redox reactions *7:* 1439, 1441
Redshift *8:* 1611, *9:* **1654-1656,** 1656 (ill.)
Reflector telescopes *10:* 1871
Refractor telescopes *10:* 1870
Reines, Frederick *10:* 1833
Reinforcement, positive and negative *9:* **1657-1659**
Reis, Johann Philipp *10:* 1867
Reitz, Bruce *10:* 1926
Relative dating *4:* 616
Relative motion *9:* 1660
Relativity, theory of *9:* **1659-1664**
Relaxation techniques *1:* 118
REM sleep *9:* 1747
Reproduction *9:* **1664-1667,** 1664 (ill.), 1666 (ill.)
Reproductive system *9:* **1667-1670,** 1669 (ill.)
Reptiles *9:* **1670-1672,** 1671 (ill.)
Reptiles, age of *8:* 1462
Respiration
Respiration *2:* 392, *9:* **1672-1677**
Respiratory system *9:* **1677-1683,** 1679 (ill.), 1682 (ill.)

Retroviruses *10:* 1978
Reye's syndrome *1:* 8
Rh factor *9:* **1683-1685,** 1684 (ill.)
Rheumatoid arthritis *1:* 183
Rhinoplasty *8:* 1527
Rhodophyta *1:* 94
Ribonucleic acid *7:* 1390, 1392 (ill.)
Rickets *6:* 1219, *7:* 1403
Riemann, Georg Friedrich Bernhard *10:* 1899
Rift valleys *7:* 1303
Ritalin *2:* 238
Rivers *9:* **1685-1690,** 1687 (ill.), 1689 (ill.)
RNA *7:* 1390, 1392 (ill.)
Robert Fulton *10:* 1835
Robotics *1:* 189, *9:* **1690-1692,** 1691 (ill.)
Robson, Mount *7:* 1357
Rock carvings and paintings *8:* 1491
Rock cycle *9:* 1705
Rockets and missiles *9:* **1693-1701,** 1695 (ill.), 1697 (ill.), 1780 (ill.)
Rocks *9:* **1701-1706,** 1701 (ill.), 1703 (ill.), 1704 (ill.)
Rocky Mountains *7:* 1301, 1357
Roentgen, William *10:* 2033
Rogers, Carl *8:* 1596
Root, Elijah King *7:* 1237
Ross Ice Shelf *1:* 149
Roundworms *8:* 1471
RR Lyrae stars *10:* 1964
RU-486 *3:* 565
Rubidium *1:* 102
Rural techno-ecosystems *2:* 302
Rush, Benjamin *9:* 1713
Rust *7:* 1442
Rutherford, Daniel *7:* 1345
Rutherford, Ernest *2:* 233, *7:* 1337

S

Sabin vaccine *8:* 1548
Sabin, Albert *8:* 1549
Sahara Desert *1:* 52
St. Helens, Mount *10:* 1996
Salicylic acid *1:* 6
Salk vaccine *8:* 1548
Salk, Jonas *8:* 1548, *10:* 1959
Salyut 1 *9:* 1781, 1788

Index

Samarium *6:* 1163
San Andreas Fault *5:* 854
Sandage, Allan *8:* 1610
Sarcodina *8:* 1592
Satellite television *10:* 1877
Satellites *9:* **1707-1708,** 1707 (ill.)
Saturn (planet) *9:* **1708-1712,** 1709 (ill.), 1710 (ill.)
Savanna *2:* 296
Savants *9:* **1712-1715**
Saxitoxin *2:* 288
Scanning Tunneling Microscopy *10:* 1939
Scaphopoda *7:* 1289
Scheele, Carl *7:* 1345, 1444
Scheele, Karl Wilhelm *3:* 525, *6:* 1032
Schiaparelli, Giovanni *7:* 1263
Schizophrenia *8:* 1596, *9:* **1716-1722,** 1718 (ill.), 1721 (ill.)
Schmidt, Maarten *8:* 1611
Scientific method *9:* **1722-1726**
Scorpions *1:* 169
Screw *6:* 1208
Scurvy *6:* 1218, *10:* 1981, 1989
Seamounts *10:* 1994
Seashore biome *2:* 301
Seasons *9:* **1726-1729,** 1726 (ill.)
Second law of motion *6:* 1171, *7:* 1235
Second law of planetary motion *7:* 1426
Second law of thermodynamics *10:* 1886
Secondary cells *2:* 270
Secondary succession *10:* 1837, *10:* 1838
The Secret of Nature Revealed *5:* 877
Sedimentary rocks *9:* 1703
Seeds *9:* **1729-1733,** 1732 (ill.)
Segré, Emilio *1:* 163, *4:* 775, *6:* 1035, *10:* 1913
Seismic waves *4:* 703
Selenium *7:* 1449
Semaphore *10:* 1864
Semiconductors *4:* 666, 734, *10:* 1910, 1910
Semi-evergreen tropical forest *2:* 298
Senility *4:* 622
Senses and perception *8:* 1482
Septicemia plague *8:* 1519
Serotonin *2:* 350
Serpentines *1:* 191

Sertürner, Friedrich *1:* 33
Set theory *9:* **1733-1735,** 1734 (ill.), 1735 (ill.)
Sexual reproduction *9:* 1666
Sexually transmitted diseases *9:* **1735-1739,** 1737 (ill.), 1738 (ill.)
Shell shock *9:* 1826
Shepard, Alan *9:* 1779
Shockley, William *10:* 1910
Shoemaker, Carolyn *6:* 1151
Shoemaker-Levy 9 (comet) *6:* 1151
Shooting stars. *See* **Meteors and meteorites**
Shumway, Norman *10:* 1926
SI system *10:* 1950
Sickle-cell anemia *2:* 320
SIDS. *See* **Sudden infant death syndrome (SIDS)**
Significance of relativity theory *9:* 1663
Silicon *2:* 400, 401
Silicon carbide *1:* 2
Silver *8:* 1566, 1569
Simpson, James Young *1:* 143
Sitter, Willem de *3:* 575
Skeletal muscles *7:* 1310 (ill.), **1311-1313**
Skeletal system *9:* **1739-1743,** 1740 (ill.), 1742 (ill.)
Skin *2:* 362
Skylab *9:* 1781, 1788
Slash-and-burn agriculture *9:* **1743-1744,** 1744 (ill.)
Sleep and sleep disorders *9:* **1745-1749,** 1748 (ill.)
Sleep apnea *9:* 1749, *10:* 1841
Slipher, Vesto Melvin *9:* 1654
Smallpox *10:* 1957
Smell *9:* **1750-1752,** 1750 (ill.)
Smoking *1:* 34, 119, *3:* 476, *9:* 1682
Smoking (food preservation) *5:* 890
Smooth muscles *7:* 1312
Snakes *9:* **1752-1756,** 1754 (ill.)
Soaps and detergents *9:* **1756-1758**
Sobrero, Ascanio *5:* 844
Sodium *1:* 100, 101 (ill.)
Sodium chloride *6:* 1096
Software *3:* 549-554
Soil *9:* **1758-1762.** 1760 (ill.)
Soil conditioners *1:* 67
Solar activity cycle *10:* 1848

Index

Solar cells *8:* 1504, 1505
Solar eclipses *4:* 724
Solar flares *10:* 1846, 1848 (ill.)
Solar power *1:* 115, 115 (ill.)
Solar system *9:* **1762-1767,** 1764 (ill.), 1766 (ill.)
Solstice *9:* 1728
Solution *9:* **1767-1770**
Somatotropic hormone *5:* 797
Sonar *1:* 22, *9:* **1770-1772**
Sørenson, Søren *8:* 1495
Sound. *See* **Acoustics**
South America *9:* **1772-1776,** 1773 (ill.), 1775 (ill.)
South Asia *1:* 197
Southeast Asia *1:* 199
Space *9:* **1776-1777**
Space probes *9:* **1783-1787,** 1785 (ill.), 1786 (ill.)
Space shuttles *9:* 1782 (ill.), 1783
Space station, international *9:* **1788-1792,** 1789 (ill.)
Space stations *9:* 1781
Space, curvature of *3:* 575, *7:* 1428
Space-filling model *7:* 1286, 1286 (ill.)
Space-time continuum *9:* 1777
Spacecraft, manned *9:* **1777-1783,** 1780 (ill.), 1782 (ill.)
Spacecraft, unmanned *9:* 1783
Specific gravity *4:* 625
Specific heat capacity *6:* 1045
Spectrometer *7:* 1239, 1240 (ill.)
Spectroscopes *9:* 1792
Spectroscopy *9:* **1792-1794,** 1792 (ill.)
Spectrum *9:* 1654, **1794-1796**
Speech *9:* **1796-1799**
Speed of light *6:* 1190
Sperm *4:* 785, *5:* 800, *9:* 1667
Spiders *1:* 169
Spina bifida *2:* 321, 321 (ill.)
Split-brain research *2:* 346
Sponges *9:* **1799-1800,** 1800 (ill.)
Sporozoa *8:* 1592
Sprengel, Christian Konrad *5:* 877
Squid *7:* 1289
Staphylococcus *2:* 258, 289
Star clusters *9:* **1808-1810,** 1808 (ill.)
Starburst galaxies *9:* **1806-1808,** 1806 (ill.)
Stars *9:* **1801-1806,** 1803 (ill.), 1804 (ill.)
  binary stars *2:* 276-278
  brown dwarf *2:* 358-359
  magnetic fields *9:* 1820
  variable stars *10:* 1963-1964
  white dwarf *10:* 2027-2028
Static electricity *4:* 742
Stationary fronts *1:* 82
Statistics *9:* **1810-1817**
Staudinger, Hermann *8:* 1565
STDs. *See* **Sexually transmitted diseases**
Steam engines *9:* **1817-1820,** 1819 (ill.)
Steel industry *6:* 1098
Stellar magnetic fields *9:* **1820-1823,** 1822 (ill.)
Sterilization *3:* 565
Stomach ulcers *4:* 656
Stone, Edward *1:* 6
Stonehenge *1:* 173, 172 (ill.)
Stoney, George Johnstone *4:* 771
Storm surges *9:* **1823-1826,** 1825 (ill.)
Storm tide *9:* 1824
Strassmann, Fritz *7:* 1361
Stratosphere *2:* 213
Streptomycin *1:* 155
Stress *9:* **1826-1828**
Strike lines *5:* 988
Stroke *2:* 350, 351
Strontium *1:* 105
Subatomic particles *10:* **1829-1834,** 1833 (ill.)
Submarine canyons *3:* 562
Submarines *10:* **1834-1836,** 1836 (ill.)
Subtropical evergreen forests *5:* 908
Succession *10:* **1837-1840,** 1839 (ill.)
Sudden infant death syndrome (SIDS) *10:* **1840-1844**
Sulfa drugs *1:* 156
Sulfur *6:* 1096, *7:* 1446
Sulfur cycle *7:* 1448, 1448 (ill.)
Sulfuric acid *7:* 1447
Sun *10:* **1844-1849,** 1847 (ill.), 1848 (ill.)
  stellar magnetic field *9:* 1821
Sun dogs *2:* 224
Sunspots *6:* 1077
Super Collider *8:* 1482
Superclusters *9:* 1809
Superconducting Super Collider *10:* 1852

Index

Superconductors *4:* 734, *10:* **1849-1852,** 1851 (ill.)
Supernova *9:* 1654, *10:* **1852-1854,** 1854 (ill.)
Supersonic flight *1:* 43
Surgery *8:* 1527-1531, *10:* **1855-1858,** 1857 (ill.), 1858 (ill.)
Swamps *10:* 2024
Swan, Joseph Wilson *6:* 1088
Symbolic logic *10:* **1859-1860**
Synchrotron *8:* 1481
Synchrotron radiation *10:* 2037
Synthesis *9:* 1648
Syphilis *9:* 1736, 1738 (ill.)
Système International d'Unités *10:* 1950
Szent-Györyi, Albert *6:* 1219

T

Tagliacozzi, Gasparo *8:* 1528
Tapeworms *8:* 1472
Tarsiers *8:* 1572, 1573 (ill.)
Tasmania *2:* 241
Taste *10:* **1861-1863,** 1861 (ill.), 1862 (ill.)
Taste buds *10:* 1861 (ill.), 1862
Tay-Sachs disease *2:* 320
TCDD *1:* 54, *4:* 668
TCP/IP *6:* 1126
Tears *5:* 852
Technetium *4:* 775, *10:* 1913
Telegraph *10:* **1863-1866**
Telephone *10:* **1866-1869,** 1867 (ill.)
Telescope *10:* **1869-1875,** 1872 (ill.), 1874 (ill.)
Television *5:* 871, *10:* **1875-1879**
Tellurium *7:* 1449, 1450
Temperate grassland *2:* 296
Temperate forests *2:* 295, *5:* 909, *8:* 1644
Temperature *6:* 1044, *10:* **1879-1882**
Terbium *6:* 1163
Terrestrial biomes *2:* 293
Testes *5:* 800, *8:* 1599, *9:* 1667
Testosterone *8:* 1599
Tetanus *2:* 258
Tetracyclines *1:* 158
Tetrahydrocannabinol *6:* 1224
Textile industry *6:* 1097

Thalamus *2:* 342
Thallium *1:* 126
THC *6:* 1224
Therapy, physical *8:* 1511-1513
Thermal energy *6:* 1044
Thermal expansion *5:* 842-843, *10:* **1883-1884,** 1883 (ill.)
Thermodynamics *10:* **1885-1887**
Thermoluminescence *4:* 618
Thermometers *10:* 1881
Thermonuclear reactions *7:* 1368
Thermoplastic *8:* 1533
Thermosetting plastics *8:* 1533
Thermosphere *2:* 213
Thiamine. See **Vitamin B1**
Third law of motion *6:* 1171
Third law of planetary motion *7:* 1426
Thomson, Benjamin *10:* 1885
Thomson, J. J. *2:* 233, *4:* 771
Thomson, William *10:* 1885, 1882
Thorium *1:* 26
Thulium *6:* 1163
Thunder *10:* 1889
Thunderstorms *10:* **1887-1890,** 1889 (ill.)
Thymus *2:* 329, *5:* 798
Thyroxine *6:* 1035
Ticks *1:* 170, *8:* 1475
Tidal and ocean thermal energy *1:* 117
Tides *1:* 117, *10:* **1890-1894,** 1892 (ill.), 1893 (ill.)
Tigers *5:* 859
Time *10:* **1894-1897,** 1896 (ill.)
Tin *2:* 401, 402
TIROS 1 *2:* 217
Titan *9:* 1711
Titania *10:* 1954
Titanic *6:* 1081
Titius, Johann *1:* 201
Tools, hand *6:* 1036
Topology *10:* **1897-1899,** 1898 (ill.), 1899 (ill.)
Tornadoes *10:* **1900-1903,** 1900 (ill.)
Torricelli, Evangelista *2:* 265
Touch *10:* **1903-1905**
Toxins, poisons and *8:* 1542-1546
Tranquilizers *10:* **1905-1908,** 1907 (ill.)
Transformers *10:* **1908-1910,** 1909 (ill.)
Transistors *10:* 1962, **1910-1913,** 1912 (ill.)

Index

Transition elements *10:* **1913-1923,** 1917 (ill.), 1920 (ill.), 1922 (ill.)
Transplants, surgical *10:* **1923-1927,** 1926 (ill.)
Transuranium elements *1:* 24
Transverse wave *10:* 2015
Tree-ring dating *4:* 619
Trees *10:* **1927-1931,** 1928 (ill.)
Trematodes *8:* 1473
Trenches, ocean *7:* 1410
Trevithick, Richard *6:* 1099
Trichomoniasis *9:* 1735
Trigonometric functions *10:* 1931
Trigonometry *10:* **1931-1933**
Triode *10:* 1961
Triton *7:* 1332
Tropical evergreen forests *5:* 908
Tropical grasslands *2:* 296
Tropical rain forests *5:* 908, *8:* 1642
Tropism *2:* 271
Troposphere *2:* 212
Trusses *2:* 356
Ts'ai Lun *8:* 1463
Tularemia *2:* 289
Tumors *10:* **1934-1937,** 1934 (ill.), 1936 (ill.)
Tundra *2:* 293
Tunneling *10:* **1937-1939,** 1937 (ill.)
Turbojets *6:* 1146
Turboprop engines *6:* 1146
Turbulent flow *1:* 40

U

U.S.S. *Nautilus* 10: *1836*
Ulcers (stomach) *4:* 656
Ultrasonics *1:* 23, *10:* **1941-1943,** 1942 (ill.)
Ultrasound *8:* 1640
Ultraviolet astronomy *10:* **1943-1946,** 1945 (ill.)
Ultraviolet radiation *4:* 765
Ultraviolet telescopes *10:* 1945
Uluru *2:* 240
Umbriel *10:* 1954
Uncertainty principle *8:* 1609
Uniformitarianism *10:* **1946-1947**
Units and standards *7:* 1265, *10:* **1948-1952**
Universe, creation of *2:* 273

Uranium *1:* 25, *7:* 1361, 1363
Uranus (planet) *10:* **1952-1955,** 1953 (ill.), 1954 (ill.)
Urban-Industrial techno-ecosystems *2:* 302
Urea *4:* 645
Urethra *5:* 841
Urine *1:* 139, *5:* 840
Urodeles *1:* 137
Ussher, James *10:* 1946

V

Vaccination. *See* **Immunization**
Vaccines *10:* **1957-1960,** 1959 (ill.)
Vacuoles *3:* 436
Vacuum *10:* **1960-1961**
Vacuum tube diode *4:* 666
Vacuum tubes *3:* 416, *10:* **1961-1963**
Vail, Alfred *10:* 1865
Van de Graaff *4:* 742 (ill.), *8:* 1475
Van de Graaff, Robert Jemison *8:* 1475
Van Helmont, Jan Baptista *2:* 337, 393, 404
Variable stars *10:* **1963-1964**
Vasectomy *3:* 565
Venereal disease *9:* 1735
Venter, J. Craig *6:* 1063
Venus (planet) *10:* **1964-1967,** 1965 (ill.), 1966 (ill.)
Vertebrates *10:* **1967-1968,** 1967 (ill.)
Vesalius, Andreas *1:* 139
Vesicles *3:* 433
Vibrations, infrasonic *1:* 18
Video disk recording *10:* 1969
Video recording *10:* **1968-1969**
Vidie, Lucien *2:* 266
Viè, Françoise *1:* 97
Vietnam War *1:* 55, *3:* 460
Virtual reality *10:* **1969-1974,** 1973 (ill.)
Viruses *10:* **1974-1981,** 1976 (ill.), 1979 (ill.)
Visible spectrum *2:* 221
Visualization *1:* 119
Vitamin A *6:* 1220, *10:* 1984
Vitamin B *10:* 1986
Vitamin B_1 *6:* 1219
Vitamin B_3 *6:* 1219
Vitamin C *6:* 1219, *10:* 1981, 1987, 1988 (ill.)

Vitamin D *6:* 1219, *10:* 1985
Vitamin E *10:* 1985
Vitamin K *10:* 1986
Vitamins *7:* 1401, *10:* **1981-1989,** 1988 (ill.)
Vitreous humor *5:* 851
Viviparous animals *2:* 317
Vivisection *10:* **1989-1992**
Volcanoes *7:* 1411, *10:* **1992-1999,** 1997 (ill.), 1998 (ill.)
Volta, Alessandro *4:* 752, *10:* 1865
Voltaic cells *3:* 437
Volume *10:* **1999-2002**
Von Graefe, Karl Ferdinand *8:* 1527
Vostok *9:* 1778
Voyager 2 *10:* 1953
Vrba, Elisabeth *1:* 32

W

Waksman, Selman *1:* 157
Wallabies, kangaroos and *6:* 1153-1157
Wallace, Alfred Russell *5:* 834
War, Peter *10:* 1924
Warfare, biological. *2:* 287-290
Warm fronts *1:* 82
Waste management *7:* 1379, *10:* **2003-2010,** 2005 (ill.), 2006 (ill.), 2008 (ill.)
Water *10:* **2010-2014,** 2013 (ill.)
Water cycle. *See* **Hydrologic cycle**
Water pollution *8:* 1556, 1561
Watson, James *3:* 473, *4:* 786, *5:* 973, 980 (ill.), 982, *7:* 1389
Watson, John B. *8:* 1595
Watt *4:* 746
Watt, James *3:* 606, *9:* 1818
Wave motion *10:* **2014-2017**
Wave theory of light *6:* 1187
Wavelength *4:* 763
Waxes *6:* 1191
Weather *3:* 608-610, *10:* 1887-1890, 1900-1903, **2017-2020,** 2017 (ill.)
Weather balloons *2:* 216 (ill.)
Weather forecasting *10:* **2020-2023,** 2021 (ill.), 2023 (ill.)
Weather, effect of El Niño on *4:* 782
Wedge *6:* 1207
Wegener, Alfred *8:* 1534
Weights and measures. *See* **Units and standards**

Welding *4:* 736
Well, Percival *8:* 1539
Wells, Horace *1:* 142
Went, Frits *6:* 1051
Wertheimer, Max *8:* 1595
Wetlands *2:* 299
Wetlands *10:* **2024-2027,** 2024 (ill.)
Whales *3:* 448
Wheatstone, Charles *10:* 1865
Wheel *6:* 1207
White blood cells *2:* 328, 1085 (ill.)
White dwarf *10:* **2027-2028,** 2027 (ill.)
Whitney, Eli *6:* 1098, *7:* 1237
Whole numbers *1:* 180
Wiles, Andrew J. *7:* 1394
Willis, Thomas *4:* 640, *9:* 1718
Wilmut, Ian *3:* 487
Wilson, Robert *8:* 1637
Wind *10:* **2028-2031,** 2030 (ill.)
Wind cells *2:* 218
Wind power *1:* 114, 114 (ill.)
Wind shear *10:* 2031
Withdrawal *1:* 35
Wöhler, Friedrich *7:* 1428
Wöhler, Hans *1:* 124
Wolves *2:* 383, 383 (ill.)
World Wide Web *6:* 1128
WORMs *3:* 533
Wright, Orville *1:* 75, 77
Wright, Wilbur *1:* 77
Wundt, Wilhelm *8:* 1594

X

X rays *4:* 764, *8:* 1639, *10:* 1855, **2033-2038,** 2035 (ill.), 2036 (ill.)
X-ray astronomy *10:* **2038-2041,** 2040 (ill.)
X-ray diffraction *4:* 650
Xanthophyta *1:* 95
Xenon *7:* 1349, 1352
Xerography *8:* 1502
Xerophthalmia *6:* 1220
Xylem *6:* 1175, *8:* 1523

Y

Yangtze River *1:* 199
Yeast *10:* **2043-2045,** 2044 (ill.)
Yellow-green algae *1:* 95

Index

Index

Yoga *1:* 119
Young, Thomas *6:* 1113
Ytterbium *6:* 1163

Z

Zeeman effect *9:* 1823
Zeeman-Doppler imaging *9:* 1823

Zehnder, L. *6:* 1116
Zeppelin, Ferdinand von *1:* 75
Zero *10:* **2047-2048**
Zoophobia *8:* 1497
Zooplankton *8:* 1521, 1522
Zosimos of Panopolis *1:* 84
Zweig, George *10:* 1829
Zworykin, Vladimir *10:* 1875
Zygote *4:* 787